RETAINING WOMEN IN TECH

Shifting the Paradigm

Karen Holtzblatt

Nicola Marsden

T0092642

Retaining Women in Tech

Shifting the Paradigm

Synthesis Lectures on Professionalism and Career Advancement for Scientists and Engineers

Editors

Charles X. Ling, *University of Western Ontario*

Qiang Yang, *Hong Kong University of Science and Technology*

Synthesis Lectures on Professionalism and Career Advancement for Scientists and Engineers includes short publications that help students, young researchers, and faculty become successful in their research careers. Topics include those that help with career advancement, such as writing grant proposals; presenting papers at conferences and in journals; social networking and giving better presentations; securing a research grant and contract; starting a company, and getting a Masters or Ph.D. degree. In addition, the series publishes lectures that help new researchers and administrators to do their jobs well, such as: how to teach and mentor, how to encourage gender diversity, and communication.

A Handbook for Analytical Writing: Keys to Strategic Thinking
William E. Winner
March 2013

© Springer Nature Switzerland AG 2022

Reprint of original edition © Morgan & Claypool 2022

All rights reserved. No part of this publication may be reproduced, stored in a retrieval system, or transmitted in any form or by any means—electronic, mechanical, photocopy, recording, or any other except for brief quotations in printed reviews, without the prior permission of the publisher.

Retaining Women in Tech: Shifting the Paradigm
Karen Holtzblatt and Nicola Marsden

ISBN: 978-3-031-79196-3 paperback
ISBN: 978-3-031-79208-3 PDF
ISBN: 978-3-031-79220-5 hardcover

DOI 10.1007/978-3-031-79208-3

A Publication in the Springer series
SYNTHESIS LECTURES ON PROFESSIONALISM AND CAREER ADVANCEMENT FOR SCIENTISTS AND ENGINEERS #5
Series Editor: Charles X. Ling, University of Western Ontario and Qiang Yang, Hong Kong University of Science and Technology

Series ISSN 2329-5058 Print   2329-5066 Electronic

Retaining Women in Tech

Shifting the Paradigm

Karen Holtzblatt
WITops and InContext Design

Nicola Marsden
Heilbronn University, Germany

SYNTHESIS LECTURES ON PROFESSIONALISM AND CAREER ADVANCEMENT FOR SCIENTISTS AND ENGINEERS #6

ABSTRACT

For over 40 years, the tech industry has been working to attract more women. Yet, women continue to be underrepresented in technology jobs compared to other professions. Worse, once hired, women leave the field mid-career twice as often as men. In 2013, Karen Holtzblatt launched The Women in Tech Retention Project at WITops.org, dedicated to understanding what helps women in tech thrive. In 2014, Nicola Marsden joined the effort, bringing her extensive knowledge and research on gender and bias for women in tech. Together with worldwide volunteers, this research identified what helps women thrive and practical interventions to improve women's experience at work.

In this book, we share women's stories, our research, relevant literature, and our perspective on making change to help retain women. All the research and solutions we share are based on deep research and user-centered ideation techniques. Part I describes the @Work Experience Framework and the six key factors that help women thrive: a dynamic valuing team; stimulating projects; the push into challenges with support; local role models; nonjudgmental flexibility to manage home/work balance; and developing personal power. Employees thinking of leaving their job have significantly lower scores on these factors showing their importance for retention.

Part II describes tested interventions that redesign work practices to better support women, diverse teams, and all team members. We chose these interventions guided by data from over 1,000 people from multiple genders, ethnicities, family situations, and countries. Interventions target key processes in tech: onboarding new hires; group critique meetings; and Scrum. Interventions also address managing interpersonal dynamics to increase valuing and decrease devaluing behaviors and techniques for teams to define, monitor, and continuously improve their culture. We conclude by describing our principles for redesigning processes with an eye toward issues important to women and diverse teams.

KEYWORDS

gender, diversity, inclusion, design, software development, human-computer interaction, organization, team, stereotypes, bias, practices

Contents

Acknowledgments

The original work by the authors shared in this book was conducted with the help of worldwide professional and student volunteers. Without their help, none of this work would have been possible. We thank the following people for their contributions:

Sravya Amancherla	Alisha Gonsalves	Monika Pröbster
Rebecca Annis	Kelly Hoffman	Sameen Qayyum
Lacey Arevalo	Naishi Jain	Chris Robeck
Emily Carlson	Vichita Jienjitlert	Nina Schloss
Kinny (Yuanqi) Chen	Bill Kules	Aditi Shankar
Lindley Dahners	Ruiqi Li	Mishi Soni
Janki Desai	Andreas Paepcke	Bradley Sutherland
Carol Farnsworth	Dhruvi Patel	Xuan Zhang
Kausalya Ganesh	Christine Pisarczyk	

In addition, we thank the participants in the research and intervention iterations as well as the Directors of the HCI Masters Programs at the University of Maryland, Georgia Tech, and Carnegie Mellon University who helped with recruitment.

We also thank those at InContext who helped get this work started, especially Shelley Wood, Kelley Wagg, and Wendy Fritzke, as well as Michael Ahmadi and Anne Weibert who worked with Nicola in the Living Labs on gender.

We especially thank Shelley Wood for being our reader and commentator on the book, Claudia Herling for her wonderful illustrations for the book cover, figures, and illustrations, and Peter Verastegui for the amazing Valuing and Jerk character poster designs.

Introduction

Technology companies[1] are reshaping the experience of every person on the planet. Their impact is well beyond product development. We do not know where this journey will take us, but we do know that women are grossly underrepresented in shaping the outcome. Companies do not exclude women intentionally. Quite the contrary—tech companies are making great efforts to hire women. Still, they struggle to achieve anywhere near 50% women in their workforce or leadership. As a result, women are not participating effectively in reshaping the future of the human experience. We simply must do better.

We are writing this book for those who want to make sure that women get equal representation in shaping the world of tomorrow. We are writing for those who want to create organizations and teams that are welcoming to everyone. We are writing for those who realize that for all our efforts to bring women into tech, the existing work environment drives them away. Women love tech work, but still, they leave the field at an alarming rate.

> *Women love tech work, but they are leaving the field*

We are writing this book to share what we have learned about what it takes to retain women in tech. We invite you to use our perspective and suggestions to try new things in your organization. Whether you are an engineer or user experience professional, a team lead or manager, an academic or administrator, a VP or director, or a human resources professional, internal coach, or consultant, you can help improve the way your diverse teams work. This book focuses on women in tech. Women are our largest underrepresented group in tech. And although we did not study other underrepresented groups explicitly, you may find ideas that may apply to them as well.

THE LACK OF WOMEN IN THE TECHNOLOGY INDUSTRY

Increasing the number of women in the technology industry is a high priority—and a longstanding problem. Huge investments are being made and have been made to increase women's enrollment in college science, technology, engineering, and mathematics (STEM) programs, to improve hiring practices, and to tackle workplace issues. Yet positions go unfilled.[56, 93] Companies are failing to find and retain women and other underrepresented populations in technology positions.[78] Moreover, this is not a Silicon Valley problem. Women are glaringly absent from tech jobs worldwide.

---

[1] We use the term "technology companies" to refer to software companies producing products, websites, social media, apps, and services. This includes IT groups within businesses supporting business processes and websites. We also include technology teams producing user-facing software in hardware products and vehicles. Last, we use the term "product" to refer to anything technology professionals create in these companies.

Women in tech represent only 25% of the technology workforce in North America,[48] 18% in the EU[24], and 15% in Australia.[51]

Despite significant efforts by companies, women continue to be underrepresented in technology companies as compared to in the overall workforce.[54] Starting in the 1960s and 1970s, the second wave of the women's movement led to increased participation by women in the workplace.[77] Women fought to be educated and hired into professions that were atypical for their gender such as medicine, law, business, politics, and engineering. These "hero" professions,[80] so-called because of their heavy workload and status, were not considered appropriate careers for women. But since the women's movement of the 1970s, women have successfully entered all these fields and more.[17] Women may not represent 50% of all workers in traditionally male professions, but the growth is steady. Not so in technology.

The numbers of women in medicine, law, and business keep growing— but not in tech

Unfortunately, the number of women in tech increased until the 1980s but then flattened.[7] Women's numbers then began dropping and have now stagnated at around 15–25%, depending on the country.[41] Despite the herculean efforts of organizations like AnitaB.org and The National Center for Women & Information Technology (NCWIT) in the United States, or the Competence Center Technology, Diversity, Equal Opportunities in Germany (kompetenzz.de), as well as corporate programs to recruit women, women are not filling the ranks of technology companies.

THE LEAKY BUCKET: BEYOND FILLING THE PIPELINE

A long-standing explanation for the lack of women in tech has been the "pipeline" problem, that not enough qualified women are available to fill the positions.[47, 61] Building the pipeline has long been the focus of human resource offices' recruiting efforts[33] and academic programs trying to produce more women to take the positions available.[96] These recruiting efforts are critical to growing the ranks of women in tech. A focus on attracting more women to tech has stimulated research and new practices. These include, for example, blind resume reviews and non-gendered job ads that do not use language that alienates women.[29] These efforts work to get more women in the door to be interviewed. But what happens after they are hired?

Recruiting gets women in the door—but then 50% leave the tech field

Unfortunately, once hired, women are less likely to stay in the tech industry than men. The quit rate for women is 50% higher than the quit rate for men. Numerous studies show that women who enter the field leave more often than men.[7, 28, 37, 67] The first 12 years in technology seem to be

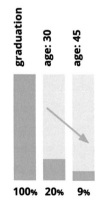

Figure 0.1: EU: Decline of women in tech [44]

the most vulnerable: 50% of women leave the field for other occupations, compared to 20% of professional women leaving non-STEM fields.[31] One sad statistic comes from the European Union (Figure 0.1): by the time they reach age 45, more than 90% of the women who graduated with a degree in information and communication technology have left the field.[44]

While men also leave the field, the number of men staying is much higher.[82] And men are more likely to turn over rather than to turn away; they take another tech job in a different firm rather than take a non-tech job.[49] Our research indicates that 45% of women in tech are thinking of leaving their job. This data is telling. More is going on with the work experience of women in technology than can be solved by better recruiting.

We are reminded of a classic quality assurance story used to illustrate a mindset that blocks finding the root cause of problems in corporate processes.

> *Imagine a janitor mopping the floor. She notices that her bucket keeps running out of water. She returns to the faucet to fill it up again and again. She does not notice the trail of water behind her. She does not see the hole in the bottom of the bucket. She is focused on refilling her pail—not finding the root cause of his empty bucket.*

Pipeline efforts are critical. But all these recruiting initiatives are for naught if women then leave the company or the industry. We need to turn our attention to the hole in the bucket, to the problem of retention.

While the numbers are disheartening given the effort to recruit women, the cost is material. Whether women leave to go to another company or abandon the field of technology altogether, companies are spending $16B per year in employee replacement.[78] When women leave, companies lose the investment they made in employee training and support. But companies also lose potential women leaders. Companies are judged by women as good places to work based on their overall number of women and percentage of women leaders. A lack of women in tech impacts cost, brand, and the ability to attract women. Worse, when women leave, it reduces the pool of potential women leaders available.

A lack of women in tech impacts budget, brand, and recruiting efforts

To achieve their stated goals, companies must focus on retention with the same energy, investment, and insight they have given to recruitment. But so far the issues of retention and advancement have been largely overlooked.[27, 60] To start we must understand why women stay in or leave technology jobs—and then how to stem the tide.

WHY DIVERSITY MATTERS

Men dominate the technology industry. But why does it matter? Tech products create the future of how we live and work. Research has definitively shown that a diverse product team improves the quality of products, breadth of innovation, success of the organization, and bottom line.[43] And companies know it. Tech has long had a diverse workforce composed of people from different parts of the world, all of whom must work together. But if we want to expand the workforce and the creativity of our teams, we need to ensure that women and other underrepresented populations can participate successfully in diverse teams.

Research over the last few decades confirms that diverse teams create more innovative solutions for complex problems than teams composed of people with similar skills and backgrounds. In his work on diversity,[70, 71] social scientist Scott Page has shown that groups with cognitive diversity consistently outperform groups selected based on a single criterion like the highest score on an aptitude test. His work and that of others[43] have time and time again shown that innovation is fostered when people with different perspectives, knowledge, experience, skills, and backgrounds are brought together to solve complex problems.

A diverse product team produces more innovative products and increases profit

Diversity within a team increases the breadth of its cognitive and behavioral repertoire. A review of research on team performance from 1960 onward also shows that teams of similar individuals deliver poorer performance when confronted with the level of complexity faced in technology companies.[13] In addition, a homogeneous team, such as an all-male team of like-minded people with similar skills and backgrounds, is more likely to ignore the needs of underrepresented populations.[10]

When women are added to the team, we also add their breadth of skills, ideas, and experience that comes from different life experiences. We enhance the chance that a superior solution will emerge and that the needs of women will be considered. So it is not surprising that diverse teams produce products that result in greater revenue and profit[36] as well as patents.[15, 88] And women CEOs also run more profitable companies.[42]

But hiring practices work against building diverse teams. People are more comfortable hiring those who have similar ways of working and thinking and come from similar backgrounds as those already on the team.[50] We call this the like-me hiring phenomenon. As we have said, like-me teams decrease diversity in skill and experience and so too the potential for creativity. But worse, like-me teams increase the probability of engaging in what is called the fundamental design error.[75] This occurs when teams define products through introspection based on their personal experience and perception of how they imagine the product will be used. When teams design from personal experience and comfort, they tend to assume their own needs and wants are similar to what other people need and want.[85] This is especially true in the absence of any rigorous data gathering from the target product populations.

A "like-me" team assumes that everyone's needs are like theirs, missing key design requirements

When team members are similar to each other this natural human myopia is boosted. Homogeneous teams are more likely to agree and reinforce initial design ideas of what makes a good product concept.[95] But if we put women and other underrepresented populations on the team, we are less likely to make silly design choices and miss product opportunities altogether.[94] For example:

- Apple's HealthKit, launched in 2014, tracked a wide range of metrics for personal health. For example, it tracked steps, stair climbing, cycling, or potassium intake. Apple claimed that it let users monitor all their most important metrics. But the product did not include a way to track women's menstrual cycles—a very important metric for 50% of the population![21]

- If a user tells Siri, Apple's private virtual assistant, that "Someone stole my computer," Siri not only understands the message but also walks the user through a process of what can be done. But if a user asks Siri about an issue more typical to women, "I've been beaten up by my husband," Siri points to a web page that says the husband most likely will do it again.[57] Siri provides no help.

- Cell phones are designed for everyone, but they fit the average man's hand much better than women's hands. As Zeynep Tufekci, the author of Twitter and Tear Gas,[91] noticed, she could not take photos above her head with one hand to document the use of tear gas—something she had seen men with larger hands do all the time.[90]

Let's not even talk about the invention of the mammogram, which led feminist human–computer interaction (HCI) researchers to explore how a similar diagnostic system for testicular exams might be received.[5] Like-me teams may think they are addressing the whole population, that they can design for everyone or for "one size fits all"; but they can't. When women are not on the team, their needs are too often not considered. As a result, products meant for everyone often fail to attract and support 50% of its target population.

What might our industry invent if we use the full potential of diversity? Companies know that to produce the most competitive and transformative products, they must form and successfully manage diverse teams. And they are spending lots of money and effort trying to make it happen. But do we really understand how to help diverse teams thrive? It's relatively easy to manage an organization when everyone is like me—not so when we are all different. With diverse teams, we have more than a project timeline or the best design to worry about. If we want the voices of women and other underrepresented populations to be heard, we must understand how to manage diverse people working together.

UNDERSTANDING RETENTION:
IF GOING TO WORK ISN'T FUN, WHY DO IT?

We spend more of our waking hours at work than anywhere else. But if the day-in-and-day-out experience of work isn't positive, why would anyone stay? We are well past the 1970s when women were opening doors to be able to work in fields they had been culturally barred from. That generation knew they wanted to do the work, wanted to change the culture, and the workplace would not be welcoming. Not so today. The current generation of women in tech may expect equality, value, and the ability to do the work they enjoy. Today, women are working in many industries, but the bad reputation of technology is not inviting them in. Tech companies compete for talent, and that means more than money, benefits, and free food. The "problem" of women in tech isn't a "woman's problem"; it's a technology work culture problem. Let's start to understand it.

IS TECH CULTURE THE RETENTION CULPRIT?

The poor retention of women in tech is especially troubling since 74% of women in tech report "loving their work".[7] But, unfortunately, too many women are not choosing or staying in technology

Retaining women in tech isn't a woman problem— It's a tech culture problem

jobs. The culture of the workplace in technology companies is often targeted as the source of the problem. Over the last 40 years, reports of gender discrimination and sexual harassment have increased (e.g. [19, 26, 46, 84]). Research into why women leave the field points to workplace factors and the overall culture as key reasons.[18, 80] Here are some persistent findings:

Stereotyping: The more stereotyping and bullying experienced, the shorter the length of time that employees remained at their previous company.[78] Fifty percent of women in STEM jobs say they have experienced gender-related discrimination in the workplace, more than women in non- STEM jobs (41%), and far more than men in STEM positions (19%).[27] Perceived role ambiguity, lack of community, unfairness, and interpersonal conflict are common complaints.[81] Women's intentions to stay on the job are lowered when they see other women undervalued, belittled, or ignored.[53]

Male-dominated culture: Tech is a male-dominated culture. Any male-dominated environment is generally associated with more harassment of women. As a result, women have higher stress and are more likely to leave the job.[69] This creates a hostile workplace for women.[18, 80] Tech makes these dynamics worse because the field of technology itself is seen as masculine.[73] So, men feel they belong in the tech field and women can feel they do not fit in.[35, 86]

Competitive culture: In a "hero" culture, the hero engineer who saves the project at the 11th hour is rewarded, not the employees who work to prevent problems and deliver on time. A competitive hero culture often contradicts the espoused company culture, which cites collaboration as a key value. This "ethos of competition" alienates women and undermines a collaborative environment.[72]

Espoused values: We like to say that tech is a meritocracy, that success is all about skill. But women's skills are perceived through the lens of gender expectations. Women's coding, for example, is evaluated more harshly.[87] Companies may declare that they are supporting diversity. But that doesn't make it so. In fact, companies that believe themselves to be the most meritocratic are often the least, because they do not take action to mitigate inevitable gender expectations.[23, 68] So, companies with an espoused value of meritocracy may ironically show greater gender bias.[14]

These findings highlight that the implicit workplace culture of technology companies creates a negative daily experience for women at work. And for women of color in tech, the negative workplace experiences are significantly worse.[45] Competition, stereotyping, ostracization, conflict, and lack of interpersonal safety are too often what women face when choosing to work in technology.[16] To retain women in tech, we must make real changes in the experience of daily life at work.

This research points to culture as the starting point for change. But culture is vague. How it plays out in behavior between people is idiosyncratic and hard to characterize. Women absolutely can experience a negative workplace, yet some women in tech do stay. We have documented the negative experience, but do we really understand what is going on in the tech workplace? Do we understand what is necessary for women to thrive? What are the work experiences of women who do thrive? To make real change, we must shift the inquiry to include understanding why women stay.

DOES FAMILY AND HOME/WORK BALANCE MATTER?

A second oft-cited problem of retaining women in tech is the issue of home/work balance. If only we had better support for family, goes the thinking, women would stay. But is it really a root cause? Another hero mindset is that tech workers need to be available 24/7 to get the product or deliverables out. Engineering culture is thought to discriminate against working women with family responsibilities who can't devote 24/7 to the job. But it turns out that other

Family issues are not the main reason women leave technology careers

hero industries such as medicine, law, and business have the same expectations for work, yet the numbers of women in those fields are higher. Every industry must do better to support a balance between family or personal life and work. But recent studies show that family issues are not the primary or the main reason women leave technology occupations.[55, 80, 83]

Introducing family-friendly policies is good; technology companies, like others, are making some headway. Some corporations are increasing parental leave, building nursing rooms, making lists of childcare workers available, and creating support programs for new parents.[59] But, of course, parenting demands last much longer than the first few months after birth. Many of these "new" approaches were tried in the 1980s at companies like Hewlett-Packard, which offered job sharing and other types of support. But we have a long way to go to a work culture that truly supports real careers for all parents as they care for their children.

More importantly, a focus on home/work balance as a "women's issue" perpetuates negative expectations and stereotyping. Declaring that family support is something that is being done "for

| *Managing family isn't a "women's issue." Men depend on home/work policy more than women* |

women" suggests that women need special treatment, that they are less capable of handling a job than men without support. And it implies that family duties affect women more than men, thereby perpetuating traditional role expectations.[1] But interestingly, it is men with family obligations who depend on family/work balance

the most. Women are more resilient; they know how to juggle their commitments.[66]

Fixing family policy is a good corporate goal. But since home/work balance is not the root cause of why women leave tech, it is not the root solution. Providing parental support is a must for every company, as we learned so well during the COVID pandemic. But fixing family policy will not be the quick fix to hiring and keeping women in tech.

THE WOMEN IN TECH RETENTION PROJECT: WHY WOMEN STAY

Recruiting women into a culture that then rejects or repels them undermines all the hard work done to increase women's interest in technology fields. But not all women leave. Let's take a new approach and ask why women thrive. What are the practices, interactions, relationships, and personal strategies that help women succeed? This is the goal of the Women in Tech Retention Project launched in 2014.

Through a series of qualitative and quantitative studies, we explored the daily work experiences of women in technology companies doing jobs at the heart of product development. These women represented a mix of races, ethnicities, ages, household types, job types, career lengths, and sexual orientations. The focus of this initial research was to understand what keeps women committed to tech work and helps them thrive. We began with in-person contextual interviews[39] focused on understanding the daily life experiences of successful women. We wanted to understand what draws and keeps them engaged at work. We explored the everyday experiences of product teams,

| *Those with lower scores on key factors are more likely to be thinking of leaving their job* |

management, evaluation, promotion, and working with others. We did not focus on issues related to recruiting, family benefits, or sexual harassment unless the women raised them.

In these qualitative interviews, we examined what contributes to success for women and if the lack of those same things nudged them to leave a job. The emergent factors that help women thrive let us explore if these are the missing experiences underlying why women leave a job. As applied researchers, we also sought to understand and find the most powerful intervention points, areas where companies might change their practice, policy, programs, or management approaches.

Through our analysis, we identified the key factors essential to why women stay in technology jobs. The factors are organized into The @Work Experience Framework discussed in Part I. Here we briefly describe them.[25, 38, 40, 58]

Dynamic, Valuing Team: Women thrive in a dynamic, work-focused team and/or partnership where they can lead, follow, feel valued, and talk about life outside of work within their team.

Stimulating Work: Women love working on challenging technical problems, products, or research questions important to the company, the industry, or the world. They switch jobs when bored.

Push & Support: Women may not feel qualified for the next challenge. But when pushed by trusted managers, colleagues, or family, they take it on and succeed—provided they have support to strategize with others, ask questions, and falter.

Local Role Models: Women need coaching relationships in their company to help them succeed. Work buddies with more experience help them navigate their careers. Senior co-workers and managers also reveal the experience of daily life after promotion. If their lives look undesirable, women may not seek promotion.

Nonjudgmental Flexibility: Women with children thrive if the team and managers flex to everyone's life commitments. These women too often feel judged for meeting home commitments. Being given flexibility by the team and managers shows them that they are valued.

Personal Power: Women can have self-doubt about their skills, readiness, and value. Self-esteem increases with positive feedback, helpful critique, clear expectations, and good coaching.

We validated these factors and further explored women's experience through a series of surveys with more than 900 respondents worldwide.[40] The resulting @Work Experience Measure is a 15-minute survey assessing employees' experiences related to the factors. This quantitative research validated the importance of these factors as necessary ingredients to a positive working environment.

One of our most telling statistics is that both women and men who reported that they were "thinking of leaving the job" scored lower on all the factors above except Nonjudgmental Flexibility, which is related to home/work balance (Figure 0.2). Given the research findings on family issues discussed above, this is not surprising.

Interestingly in our quest for factors of workplace experience that matter to women, we also found the work experiences that matter to all. But clearly since so many more women leave the field than men, the tech work culture is not delivering these critical work experiences to women as reliably as to men.

Armed with key factors we now have a practical starting place to develop a strategy to guide organizational change. Change is never easy but having a direction makes it possible.

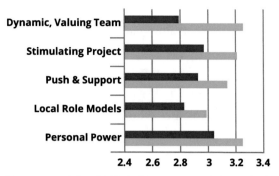

Figure 0.2: The @Work Experience Factors: Differences between people who are thinking about leaving their job (purple) vs. not (green).[40]

PRACTICAL INTERVENTIONS THAT WORK

Wouldn't it be great if people could just figure out how to work together smoothly? Everyone is looking for the secret sauce, the quick fix, the policy change, or the workshop that would make the tough task of diverse people working together go away—or at least feel approachable. We are often asked, why is the technology industry different from the other hero professions? Certainly, these professions have their challenges and will benefit from many insights of our work. We hope so. But these industries are not failing to attract and retain women in the same way as tech. What's the difference?

Technology is a "maker culture." In a maker culture, people of different job roles must collaborate and coordinate continuously. A maker culture is also a continuous feedback culture where co-workers, customers, and the response of the market evaluates the work of the team and

Tech is a maker culture with different teamwork demands than other professions

its members daily. A maker culture means people must work nose-to-nose with others to understand customer needs, invent, design, critique, code, and get buy-in for whatever they are making. This level of collaboration and feedback may characterize the bulk of a maker's workday and week. A maker culture depends on getting along with others and being in a team in ways that are simply not required in other professions.

The demand for smooth, fast collaboration may be one of the reasons that the exclusion processes are often found in maker cultures.[4] Intense interaction comes more easily with people that are like me. As a result, maker cultures run the risk of marginalizing outgroups like women and other diverse people. Marginalization undermines the dynamic collaboration necessary to include diverse perspectives on a team. But a maker situation is exactly the kind of complex challenge that benefits from diversity.

To perform well, people need to feel valued both as members of the team and for their individual characteristics, behaviors, and contributions.[76] Ostracization effectively undermines women's sense of belonging and value that is key to thriving at work. So, it is not surprising that our research shows that women thrive in technology if they experience a dynamic, valuing team. More importantly, when people report having positive experiences working with their team, they are also less likely to be thinking of leaving. See Figure 0.3.

People with positive team experiences are less likely to be thinking of leaving their jobs

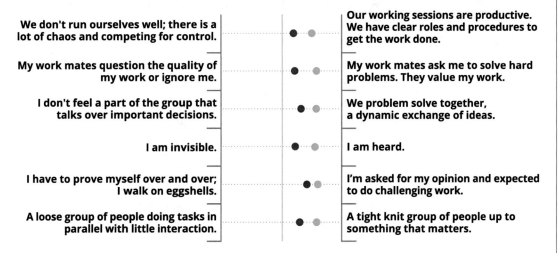

We don't run ourselves well; there is a lot of chaos and competing for control.	Our working sessions are productive. We have clear roles and procedures to get the work done.
My work mates question the quality of my work or ignore me.	My work mates ask me to solve hard problems. They value my work.
I don't feel a part of the group that talks over important decisions.	We problem solve together, a dynamic exchange of ideas.
I am invisible.	I am heard.
I have to prove myself over and over; I walk on eggshells.	I'm asked for my opinion and expected to do challenging work.
A loose group of people doing tasks in parallel with little interaction.	A tight knit group of people up to something that matters.

Figure 0.3: Items of the factor "Tight Cohesive Team" showing the differences between people who are thinking about leaving their job (purple) and those who are not (green).

Any maker culture is heavily team based and interpersonally challenging. It requires that people get along professionally while innovating in a fast-paced, dynamic creative environment. This is a challenge for any group of people. Now add diversity and the challenge increases. We need effective diverse teams in technology. We need diversity for great solutions. We need to grow a diverse workforce. And we want everyone, including women, to pursue the tech work they love. But the more diverse the people, the greater the challenge. So, what do we do?

INSANITY IS DOING THE SAME THING OVER AND OVER BUT EXPECTING A DIFFERENT RESULT

At the Grace Hopper Conference in 2018, 20,000 women in technology with diverse backgrounds came together in one room for the opening keynote. The stadium, indeed, the whole city of Houston, Texas, was filled with women who work in and want to work in technology. Female speakers on the stage gave inspiring talks to show what is possible as a woman in tech. Women led workshops, gave talks, and showed their leadership. It was a grand celebration of women—and it communicated that no woman in tech is alone anymore.

The Anita Borg Institute was founded in 1987 to bring technical women together to discuss issues they experienced at work and share resources. Anita Borg, Telle Whitney, Grace Hopper—all groundbreaking women in tech—inspired forums for women technologists to feel their power and support each other. Since then, women's groups have been created in cities where women in tech work and within many companies. The notion is that when women gather, they can gain support and perspective from others who share their challenges and experience. Networking groups constitute one of the first interventions developed to encourage women's success in tech.

Years of research into women's issues, gender roles, and bias have repeatedly revealed that socialization in families, school, the workplace, and countries defines and enforces gender differences and opportunities. To make change for women gender bias was attacked with training and workshops that have been delivered within and outside of businesses for 40 years.[20] Assertiveness training and other programs to enhance self-esteem and confidence have also grown. Professional development programs have been offered to promote women's leadership.[6] Training and workshops for women in tech include self-awareness, self-efficacy, personal visions, developing coaching relationships, and more.[92]

We have been using networking, bias training, and self-development workshops to try to change the culture and people for decades. But today, for example, women in tech are still judged to be less competent than men and the quality of their work is judged as poorer.[64] Most would vehemently deny that we judge the same work differently based on whether it was created by a woman or a man. Many people think we are beyond this kind of bias but unfortunately, we are not. Gender stereotypes and bias remain with us unconsciously and affect our judgment and choices.

Recent research into stereotyping shows that people discriminate without awareness or conscious control. This is the reason why modern assessments of stereotyping measure response time or count mistakes made under time pressure to uncover implicit associations with gender. For example, if people find a word like "warm" faster when it follows the picture of a woman than when it follows a picture of a man, this is taken as an indicator that woman are associated with warmth. A quicker response shows that we have an internalized bias that women are expected to be interpersonally warmer than men. Tests of implicit associations show that gender shapes our judgment about a person's character, abilities, and potential.[9, 34] And these automatic gender stereotypes are something we all share. Women as well as men judge women to be less competent.[8] Unconscious bias means we are all unaware of how we may stereotype others.

Even with years of effort and training, bias and stereotyping persists

Learning about implicit bias and providing professional development for women is important. Awareness allows people to better monitor their behavior to align it with their values. Awareness and the development of individual skill are necessary. Unfortunately, women's sense of belonging and feelings of discrimination are largely unaffected by whether their departments complete a diversity training.[64] Changing unconscious behavior is hard—just changing ourselves even when we want to is hard. Now add the interpersonal demands of a maker culture. It is not surprising that these approaches have had little effect.

Networking, mentoring, awareness workshops, and training have been tried for years. They helped us become aware of gender bias as a society. But these approaches are clearly not enough to radically shift company culture or the culture of everyday work. With all the

> *Women report discrimination even after their group has diversity training*

effort and the goodwill of good people, the number of women in tech continues to flatten and drop. Women are walking away from our field.

Insanity is doing the same thing over and over and expecting a different result. It's time to step back and develop new approaches.

STRUCTURE FOR FREEDOM: PRINCIPLES AND PRACTICES, NOT PERSONALITIES

William Edwards Deming, a well-known quality guru, told managers that the key to quality products is to manage their processes. Six Sigma, Agile, user-centered design, and more are based on a philosophy that if companies want a quality product, they must manage how they do the work. The idea of focusing on managing work processes and practices instead of individual performance is a radical departure for most people. Focusing on processes, practices, and procedures is particularly hard when we are prone to see the problem as people's behavior. When we interact with our team, our manager, or a person in the company, we might feel valued or devalued—welcomed or pushed away. When we interact with people, it feels personal. So, it's hard to imagine that we can effect change by focusing on how we do the work instead of the personalities and behaviors of those we interact with.

RESTRUCTURE THE PROCESS TO CHANGE BEHAVIOR

Researchers and educators have found that when women are working with or surrounded by a majority of men, their performance is impaired. One early intervention to change women's participation was to create or reinvent all-women schools for the expressed purpose of fostering women's leadership. In a mixed-gender environment, young women did not naturally assume leadership roles or speak up as did men. But in an all-female setting where no men were around to assume these roles, women took the lead. "Students have the same classes but being in a setting with only women changes their behavior. They take on leadership roles, speak up, challenge each other" says Juliane

Siegeris, professor of software engineering, who established the women-only computer science program at the University of Applied Sciences in Berlin.[79]

The women in all-female schools naturally change themselves because the school deliberately changes the situation they are in. Karen adopted this approach in the early 1990s when coaching a product team to use Contextual Design, a well-known user-centered design process.[39] Karen explains:

> We were coaching a team with six men and four women. The men were jumping into the ideation process. The women were silent or asking the men how "we" should solve things. Rather than tell anyone to change themselves or participate differently, we changed the setting. We split the team into parallel visioning sub-groups, giving the women a group of their own. The women then started throwing out ideas, participating, and inventing. Afterward, we talked to the women about what happened. When we put the team together again everyone participated.

Deliberately installing same-gender teams is a good example of how changing the environment changes behavior. We are not saying that technology companies should create all-male and all-female teams—although a little experimentation might be warranted. And we know that this kind of segregation can foster unequal opportunities and unhealthy competition. Our point here is that to change behavior we don't have to ask everyone in an organization to become unbiased. We just need to alter the context in which people interact, work together, and make decisions.

The use of blind reviews is another example of changing the situation to change the result. When the gender or race of candidates for a job are known, reviewers are inordinately influenced by their biases. Identical papers or resumes are rated higher when they are labeled with a men's name versus a women's name.[11, 63, 65, 74] The current best practice is a blind review for job applicants or academic papers, resulting in more women applicants interviewed and papers accepted.[52, 97] Without a picture of the applicant or their name revealing gender, the reviewers focus on skills, experience, and quality.

A blind audition when hiring musicians for orchestras is a powerful example of the power of changing a process to eliminate bias. Orchestras ask candidates to play behind a screen so that the jury cannot see their gender. The effect is an eye-opener. Not knowing gender heightened women's chances to be advanced by 50%.[32] The hiring jury is trying to hire the best musician for the orchestra. But gender bias plays a role when the jury can see the musicians. Eliminating that bias is easy with a simple change in the process.

To interrupt biased behavior, change the work context practice

"Stop simple diversity training focused on raising awareness," appeals Iris Bohnet.[12] Her work as a behavioral economist champions "gender equality by design"—reducing the impact of unintentional bias by redesigning the process. For example, Bohnet describes the results of her research into volunteer behavior. In same-gender groups, women and men are equally likely to volunteer. When a team is made up of different genders, the situation changes. Women volunteer more than men. Women spend more of their time doing things that the whole team benefits from, often to the detriment of their own career advancement. But when everyone is rewarded for volunteering for these team chores, men volunteer more often.[12] Rewarding the desired behavior reduces gender differences.

The authors routinely help manage diverse teams by redesigning the processes that those teams use to get the work done. In this example, Nicola shares her experience with project planning in software engineering teams.

> *Estimates of how long work will take have a direct impact on the ship date and the profit of a project. Teams we have observed frequently sabotage themselves by agreeing too fast. They may not listen to the one person who has it right. Or they may take a mental shortcut and follow the person who is seen as the leader or the expert. So, to be sure we heard every person's voice and get more reliable estimates we instituted planning poker.*

Planning poker is a method popularized in Scrum. It asks each team member to privately create their own time estimates from a sequence of values expressing their hunch of how much work the team needs to do for a given plan. First, individual team members complete their private estimates. Only then do they share it with the team. As in a poker game, the team members only reveal their cards when asked. This approach ensures that everyone participates and that outlying thinkers are not swayed by the initial majority.[62] The result is not only a better estimate but also a change in the experience. As Sylva told us:

> *I used to always let others talk when it came to estimating workload. I thought they knew better because most of them had more experience. With planning poker, I'm forced to give my own estimate. It starts important conversations, and we get better results that way. Sylva, Software Developer*

Changing the process changes behavior. This is not a special approach for diversity and gender issues. We redesign our processes to achieve a better result all the time. If we know the kinds of interactions and behavior that will help women thrive, we can design our way to diverse teams that work for everyone. Like Bohnet, the authors maintain that creating solutions to retain women in tech is a design challenge. And like any design problem, we start by gathering data to understand what is going on. Then using that data, we create and iterate practical solutions that work for real tech teams working together. Changing people is hard. But redesigning practices to interrupt bias is possible.

UNDERSTANDING TEAMS

At the center of a maker culture is the team. As we have said, how women experience daily work with their teams is critical to retention. So how diverse teams work, how good teams work, is something we must understand. Only then can we consider redesign options.

When people come together to solve a problem, it takes them a while to organize themselves to be productive. Both research and anecdotal experience from people who have lived through team formation suggest it is not always easy. Some teams succeed and some do not. Existing models of team dynamics all expect turbulences at some point. Bruce Tuckman has identified key phases in team development. Before becoming productive they go through what Tuckman calls forming, storming, norming, and then performing.[89] In other words, all teams take some time to figure out how to work together before they work together smoothly.

All teams have turbulence before figuring out how to work together—then they get productive

Connie Gersick further found that teams do not display a steady curve of performance. Rather, the length of the lifetime of a team influences when teams become productive. In the beginning, teams do very little, they engage in status seeking, procrastinating, and self-organizing. Then at the temporal midpoint, the team really starts being productive. Give a team seven days to do a task, they get started in earnest on Day 4. Give a team 12 weeks, they get started in earnest in Week 6.[30] Given an end date, a deadline, or a shipping date the midpoint of a project is the moment that is galvanizing and leads to a burst of activity in the team (Figure 0.4). Agile practices and their strict focus on timing make use of this team phenomenon when they define short sprints to move the work along.[62] Again, by changing the process, in this case the length of time, we get a different result from people.

In the quest for understanding what makes a good team work smoothly, Google's analytics team spent over two years investigating what makes Google teams great.[22] Google found that successful teams had formed a set of roles defining each person's responsibilities, a set of values related to how to get the work done, and a set of norms for interacting. The goal of every product team is clear: define, validate, test, and ship a product. But successful teams then defined how they worked together and the process they used. Some of Google's successful teams were more leader-dominated and some were more collaborative. In the end, the exact roles, values, behaviors, and process to get the work done didn't matter. Having roles, values, behaviors, and procedures that all agreed on did.

The team culture, the team's roles, values, accepted behaviors, and ways of working, were co-created and adhered to by all team members. Having a goal and knowing how to work together

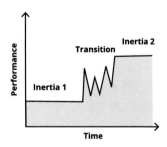

Figure 0.4: The work pace of teams: The more time you give a team the longer they take to be productive.[30]

helped the team be productive. What about diversity? Google researchers found that the group's norms typically override individual proclivities and encourage deference to the team. Teams that have a defined team way of working performed better irrespective of their diversity. Team culture creates a small social microcosm. People within this social group, no matter how different the individuals, both adjust to and co-create the way the team functions.

Diverse teams work well when they have clear roles, values, processes, and accepted ways of interacting

In other words, when teams have clear roles, values, accepted behaviors, and processes to get the work done, people collaborate productively. When they don't, they have conflict. So, what if we did it on purpose? What if the team culture and way of working were well defined and known by the members? What if we deliberately defined how we work together? Could we skip some of the turbulence? If team culture and process trump our biases, can we manage diverse teams by defining key processes, goals, roles, and norms?

Jen, a software development manager we interviewed, gives us a hint at an answer. She merged two Scrum teams that each had lost people. Both teams were working on the same larger product and were using standard Scrum processes. Jen knew Tuckman's research on teams, that any change in the team configuration, with people joining and leaving the group, will send the team back to storming to redefine how to work together. Here's what happened:

> *I expected the teams to go through a period of storming before they became productive. I called in the Agile coach, who explained that because the configuration of the group changed, they should expect turbulence. But surprisingly it didn't happen. Instead, the two teams each pulled out their Agile Team Manifestos, which state how the team agreed to work together. They examined both documents, looked for overlap, discussed any conflicts of values or behaviors, and created a new Manifesto that they all agreed to. Then they redistributed the roles and got down to work. They jockeyed a little for a week before they were off and running productively.*
> *Jen, Software Development Manager*

Why did this team adjust so fast? Because everyone on the team knew what to do. They knew their role within the project, the skill they brought to the team, the processes they were to use, and the values of how to work together. In other words, they had a clear team culture. So often when we think about culture, we pay attention to corporate culture. We don't pay attention to the team culture, the implicit rules, roles, processes, and agreements on a team. But that is what makes or breaks working together.

Corporate culture is ephemeral. We know it can affect the everyday way of working and how people treat each other. And we also know that espousing a culture of equality, respect, and collaboration doesn't make it so. Corporate culture is hard to change, but team culture is much easier.

In any maker industry, nothing will get made unless there are clear roles, processes, and expectations for collaboration. In tech, we routinely change, adopt, and adapt processes to increase the quality of what we make. Over the last 30 years tech has redesigned product development processes

Agile techniques like a Team Manifesto can help teams articulate their team culture

with practices like user-centered design, design thinking, Agile, and quality practices. We know how to do this—we just need to find the right techniques and practices to transform the experience of diverse teams. Then we can reap the value promised by diversity and retain women in tech.

FINDING CRITICAL INTERVENTION POINTS

The everyday work life of diverse teams contains many practices, interactions, relationships, values, and attitudes. We have already said that changing attitudes by increasing awareness of bias does not by itself improve the daily experience of women in technology companies. We have also said that effective interventions influence how we work; effective interventions do not target individual people for change. By adopting new practices informed by gender issues we can interrupt bias. We can ensure that all voices on a diverse team are heard and that women thrive. When all members of diverse teams participate fully, when their perspectives inform product design, we foster more successful products. For this, the whole diverse team must work well together. So, effective interventions to help women thrive must also work for the whole team. Given this perspective, what practices should we target for intervention?

Our approach to identifying the most impactful intervention points always starts with in-depth qualitative research to identify key themes. The qualitative and quantitative research that led to The @Work Experience factors pointed us to critical experiences necessary for women to thrive in the workplace.[25, 38, 40] These findings identified what women need but also revealed processes and experiences to target for additional research projects. Data from all projects provided the context for ideation to derive specific intervention recommendations: redesigned work practices, new technology support, games, workshops, and fun representations of interpersonal dynamics. We honed these intervention ideas by working with managers and people on diverse teams to try out and iterate the interventions. In-depth experience data tells us where to direct ideation; iteration with real working teams tells us whether our intervention ideas will work.

The stories of women's lives that we share throughout the book come from in-depth interviews with 150 participants and related survey data with over 1000 participants across multiple projects. The overall population of our collective research represents a wide variety of people from around the world and from many types of technology organizations. See the high-level description of the participant demographics.

Our WIT Research: Participant Demographics

Throughout the book we refer to people we have interviewed and results from our surveys. Below is a description of their high-level demographics across all our Women in Tech (WIT) research efforts related to women's experience in technology companies. To validate the factors through a survey we paid for a panel of participants. All other survey participants were worldwide volunteers.

Participants: Across all research projects we used Contextual Inquiry techniques to interview 150 people. We also surveyed over 1,000 participants.

Gender: Approximately two thirds of all participants were women, one-third were men, and 1% were non-binary.

Sexual Orientation: Of all participants, approximately 10% identified as LGBTQ+.

Ethnicity: We did not explicitly sample for ethnicities or race. But because our survey data was worldwide it includes multiple ethnicities About two thirds identified as White/Caucasian, 10% Asian Indians, 5% Black/African Americans, 5% Hispanic/Latino, and 10% with other ethnic groups. Our qualitative data also includes people from many ethnic backgrounds whether they were born in the United States or elsewhere.

Age and Marital Status: Participants were primarily between 25 and 50 years old. One fifth were single, two fifths lived with a partner (but without children), and two fifths lived with children.

Geographic Location: The majority of the participants lived in the United States. In addition, approximately one third lived in Germany and 10% lived in other countries, primarily India, the UK, Austria, Switzerland, Canada, and Mexico. We also asked our participants whether the country they lived in was their country of origin. About a quarter said that they originally came from another country. So even though our participants were weighted toward current U.S. residency, they represent multiple countries of origin.

Organizations: All participants worked in tech. Almost half of them worked for companies that are known for making software, 10% worked in consulting and professional services, 5% worked in higher education, and the remaining participants worked in other branches of tech or in tech divisions of other industries such as automotive or retail.

Job Types: For most research projects, participants were recruited to be developers, designers, user researchers, product managers, project managers, or people managers. We also had a smattering of people from other technology jobs. All participants were involved with teams making software products, systems, websites, or related work products.

The projects we draw upon include The @Work Experience Framework research; the Career Power board game development that articulates work situations and career challenges; the new hire and onboarding research leading to The Team Onboarding Checklist; The Valuing and Jerk Project to identify key behaviors that impact interpersonal relationships; and The Remote Work Project during the COVID-19 pandemic to understand the overall impact of remote working in diverse teams. We also draw upon data from Nicola's team using a living lab approach to work with women in German technology companies over a 3-year period.[2]

Some of the interventions we share with you have emerged directly from this research. Some research findings pointed us to the importance of practices the authors have developed in our collective experience consulting with diverse technology teams over many years. All the interventions we share are designed to be practical. Practical interventions are those which can be implemented and used by everyday team members and managers.

We have learned that "boiling the ocean," trying to change the whole corporate culture and each person in it from the top to the bottom, is not a reasonable goal. Instead, we look for techniques that can be easily adopted by managers and team members to start making changes to improve how diverse teams work in their organizations.

The interventions we describe fall into three areas of impact: key practices, interpersonal dynamics, and team culture. (See the diagram of critical intervention points.) Redesigning key practices that impact women's experience at work can clarify expectations, provide information needed for success, and disrupt bias. We cover interventions for team onboarding, how to run a critique meeting, and ways to tune Scrum so it ensures all voices on a diverse team are heard.

Our second area of intervention addresses interpersonal dynamics, one of the most cited issues for women in tech. The Valuing and Jerk Project identifies specific behaviors which stimulate the feeling of value and those which undermine it. We introduce the valuing and jerk character posters and how to use them in a team workshop and 1-1 interactions.

Last, we introduce interventions to define and continuously monitor team culture which too often undermines women's experience at work. The Team Manifesto helps teams form a shared understanding

2   Nicola's team used a living lab approach[2-3] to understand and create interventions to improve women's experience in technology companies. A synthesis of interview transcripts, field notes, and notes from informal interactions informed our analysis of Agile techniques with people in a variety of roles and ages. It was partially funded by the German Federal Ministry of Education and Research (BMBF), grant numbers 01FP1603 and 01FP1618. The responsibility for all content supplied lies with the authors.

of how they want to work and interact. The Process Check helps the team monitor and reflect on how well they are working together.

As an industry, we are poised for change. Tech companies know they want and need women. They need diverse people to fulfill jobs and ensure innovation. Women love tech work but will not stay unless we provide the kind of daily work life they need to thrive and excel. Our goal is to put knowledge and tools in the hands of everyone so we can influence our industry together. These interventions, and the associated principles of redesigning processes, give you a place to start redesigning your organizations.

OVERVIEW OF THE BOOK

This book shares our insights and perspective on how to better retain women in tech. Through our research, we have identified what women need to thrive and key intervention techniques to help improve the daily work practices of diverse teams. We share our findings and our interventions to help you address retention in your organizations. We also share academic research to provide additional perspective.

The structure of the book starts with this Introduction laying out the challenges of retaining women in tech, our work, and our approach to change. The remainder of the book is structured as follows.

Part I describes the six factors of The @Work Experience Framework that we have identified as essential for women to stay in technology jobs.

> **Chapter 1: A Dynamic, Valuing Team That's Up to Something Big.** We describe what women are looking for in their teams and what gets in the way of team cohesion. We look at the experience of being a dynamic team and its primacy of the team for retention.

> **Chapter 2: Stimulating Work.** We discuss women's need for challenges and what they consider to be stimulating work. We raise up the role of bias in work assignments and perceptions of women's technical competency. We emphasize the role of boredom in retention.

> **Chapter 3: The Push and Support.** We describe the importance of pushing women into challenges that they may hesitate to take on. We also emphasize that to be successful in that challenge, women need support from their co-workers and managers through conversation, coaching, and tolerance of failure. Here we address the role of challenge for retention.

Chapter 4: Local Role Models. We emphasize the role of local senior people in developing effective successful professional women. We discuss women's perception of job promotion and how the behavior and lives of senior people impact whether women wish to advance. We emphasize the need for effective coaching support for retention.

Chapter 5: Nonjudgmental Flexibility for Family Commitments. We explore the role of the team in communicating their willingness to support family obligations. We emphasize how the team's flexibility counteracts women's feelings of being judged negatively for taking time to care for children.

Chapter 6: Personal Power. We explore the way women build self-confidence and counteract their self-doubt through interactions with co-workers. Confidence allows women to participate fully in their team and contribute value. With confidence, women build their sense of Personal Power and are less likely to leave the field.

Part II describes recommended interventions that affect daily work practices, interpersonal dynamics, and team culture. We introduce each intervention and the issues for women that it addresses. We provide practical guidance on how to use each intervention and links to where support materials may be downloaded.

Chapter 7: Team Onboarding. We share the needs of new hires, the timeframes managers should consider, and the eight building blocks of connection and success that form the basis of the Team Onboarding Checklist. We share the checklist and how to use it.

Chapter 8: The Critique Meeting. We explore the challenge of creating a feedback culture and issues for women receiving critique. We share a critique meeting process and the principles behind it. We provide clear steps for running the meeting.

Chapter 9: Sneak Attacks on Key Processes: Agile. We introduce the importance of explicit practices for improving and managing women's participation and success. We highlight how making practices explicit also helps the team's overall success. Using the Agile practice of Scrum as an example, we share The Analysis Matrix, our tool to help identify where practices, values, and expectations are implicit. Using Scrum, we provide examples of how to tune a practice to be more explicit.

Chapter 10: Valuing and Jerk Behaviors. We define the key valuing and jerk behaviors and their relative value to women and men and the character posters created that represent them. We describe a workshop using the posters to help individuals and teams increase valuing and manage devaluing behaviors within the team.

Chapter 11: Building Resilience: Team Manifesto and Process Checks. We introduce the need for explicit values, expectations, professional behaviors, and a code of conduct to build a team culture. But for real impact, the team needs to become reflective and committed to continuous improvement of how they collaborate and work. These techniques help build resilient teams.

We end the book with the **Conclusion: Principles of Process Intervention for Retaining Women in Tech**. In this chapter, we reflect upon the principles we use to guide creating interventions and redesigning the practices to work better for all members of diverse teams. We also bring together high-level implications of The Remote Project for women's experience.

Throughout the book, we invite you to use our knowledge and principles to invent interventions that will work for your situation and organization.

PART I

Introduction:
The @Work Experience Framework

To retain women in tech we need to know what helps them thrive. We need to know what makes working in technology compelling for them. We need to know what women need within the team and in daily work experience so that we can create these conditions. And, we need to remember that diverse teams produce more successful and innovative products. As we have established in the Introduction, it is in the best interest of corporations to ensure that women want to work in the technology industry and are retained.

Our research focus answered the question "What do women in tech need to thrive?" We did not ask why women leave. Instead, we tried to understand what draws them to a technology job. Then we could find out if the absence of these experiences drove them away. As we said in the Introduction, women thinking of leaving their jobs scored lower on nearly all of the factors in The @Work Experience Framework. As we said, this formative research led us to other related research questions such as new hire success, an inquiry into valuing and devaluing behaviors, and the impact of remote working. We also integrate Nicola's longitudinal study of women's experience in German tech companies which includes a focus on Agile teams. All of this qualitative and quantitative research provides us with a deep understanding of the women's experience at work and what they need.

In the following chapters, we introduce each of the six factors of The @Work Experience Framework which taken together describes what women need to thrive in their daily work life. Throughout Part I we share the voices of the women we spoke to. We use their stories to introduce the experiences that help women succeed and thrive. The stories are real but the names have been changed to maintain their anonymity. We also share related findings from our survey data and relevant literature to provide a wider context for understanding women's experiences. By sharing these understandings, we hope to provide insights that will guide you when thinking about your own organizations.

As you read, think about how each factor plays out in your organization. We are not saying that the work experiences described here are only important for women in technology companies. Women in non-tech industries may also need these experiences in their work life as appropriate to those industries—but our research does not extend to this population. Nor are we saying that the need for these experiences is unique to women. For the most part, our data shows that men also look for many of these work experiences. But women do not always get them. Given that women are often in the minority in their organizations, we need to be sure that we deliberately provide the daily work life and culture described in these chapters. Only then can we ensure that women stay in our tech industry.

Each chapter introduces one of the six key factors summarized here.

Chapter 1: A Dynamic, Valuing Team That's Up to Something Big. We describe what women are looking for in their teams and what gets in the way of team cohesion. We look at the experience of being a dynamic team and its primacy of the team for retention.

Chapter 2: Stimulating Work. We discuss women's need for challenges and what they consider to be stimulating work. We raise up the role of bias in work assignments and perceptions of women's technical competency. We emphasize the role of boredom in retention.

Chapter 3: The Push and Support. We describe the importance of pushing women into challenges that they may hesitate to take on. We also emphasize that to be successful in that challenge, women need support from their co-workers and managers through conversation, coaching, and tolerance of failure. Here we address the role of challenge for retention.

Chapter 4: Local Role Models. We emphasize the role of local senior people in developing effective successful professional women. We discuss women's perception of job promotion and how the behavior and lives of senior people impact whether women wish to advance. We emphasize the need for effective coaching support for retention.

Chapter 5: Nonjudgmental Flexibility. We explore the role of the team in communicating their willingness to support family obligations. We emphasize how the team's flexibility counteracts women's feelings of being judged negatively for taking time to care for children.

Chapter 6: Personal Power. We explore the way women build self-confidence and counteract their self-doubt through interactions with co-workers. Confidence allows women to participate fully in their team and contribute value. With confidence, women build their sense of Personal Power and are less likely to leave the field.

CHAPTER 1

A Dynamic, Valuing Team That's Up to Something Big

The best team I ever worked on was when I was a coder. I worked with the same couple of coders, a product manager and the UX person, for a few years. It was me and the UX woman and the rest were guys. But gender didn't matter. We just clicked and every day was exciting. Robin, Engineering Manager

Women thrive in a dynamic, work-focused team where together they make or explore something that matters. This team culture is results-focused, no drama, non-judgmental, and punctuated with a lively exchange of ideas among high-performing, mutually respected people. Within the team women both lead and follow, feel valued for their work, and are asked for suggestions. They know what to do and how to do it. In other words, they feel connected and engaged; their skill and contribution are valued.

The Dynamic, Valuing Team factor is one of the most important factors for retention. When people we surveyed report positive experiences working with their team, they are also less likely to be thinking of leaving their job. The everyday team experience is critical for women to thrive—and for retention. Let's unpack what is going on.

SOCIAL CONNECTION IS A MUST FOR EVERYONE

I was excited to find out that I knew someone on the management team when I came to take over this group. Helga, Senior Development Manager

Human Resources evaluations of engagement have long focused on whether or not people report "having a friend" at work.[15] Our own research supports the importance of simple, social connection. Like Helga, numerous women we talked to looked for friends at work: people they know, of similar ages, from similar locations, with similar family situations, doing similar jobs, and who have similar interests. For example, Jane was the only member of her team at her location, but she wasn't lonely because she had local guys on the team who loved to play games like she did.

I'm a gamer so even though my team is remote I can connect to the guys who are local because we all love games. Jane, Software Developer

Sue hung out with another professor after work. They were both working—but in parallel—which gave them a sense of connection.

I live alone and so does this other (male) professor. We hang out after hours working side-by-side on our own stuff. This co-working lets us socialize. It's great! Anya, Assistant Professor

But being remote makes it harder to connect to each other socially the way we did when working face-to-face. Ginny feels lonely even though she is home with her husband, kids, and pets—and she is in remote meetings all day.

I am on a global remote team. Before the pandemic, I was the only person on the team in my location. But I sat with a set of women who had similar jobs on related products. Every day we'd go for coffee, take walks, have lunch, and catch up about our lives and shared work challenges. Now I never talk to these women. The only chit-chat I get is at the beginning of a team meeting for a few minutes. It's interesting to learn more about my developer team in India, but it's not the same thing as checking in with my women colleagues.
Ginny, Technical Product Manager and Product Owner

Having friends at work gives us the psychological sense of community we need as social beings. Many remote groups are encouraging chit-chat at the beginning of meetings. But this group "cocktail party" talk does not substitute for sharing our work and home lives with friends at work who both match our life stage or interests as Ginny did when physically at work. Our interviews with remote workers during the pandemic tell us that dropping in and chit-chat simply disappear when everyone on the team is remote. Very few people we talked to set up a time to connect informally with others. Ginny felt that setting up time just to talk was using work time inappropriately even though it was OK to do when co-located. Being remote changes the opportunity and the easy ability to connect with people socially. People might get the work done, but remote working can undermine the psychological sense of community at work.

Friends at work are genuine relationships between people who do similar work but who also share similar life experiences and interests. This psychological sense of community is required for people to have an overall sense of well-being. But this is not the same as the psychological sense of team required for a Dynamic Valuing Team experience.

THE PSYCHOLOGICAL SENSE OF TEAM

Behind every successful product shipment, creation of a company, or delivery of the next version of a product is the feeling of being a part of a dynamic group up to something that matters. For many, there is nothing more compelling than the kind of belonging that happens when a group of makers pulls together, uses all their skills, overcomes obstacles, brainstorms solutions, invents processes, figures out a plan, and executes successfully. Working in these dynamic teams is at the root of work engagement and satisfaction.[49] Women want this—just like men. But women are less likely to have this experience when working in tech companies populated by a majority of men.

The Dynamic Valuing Team factor implies a way of working with the set of people we interact with every day to achieve the team's goal. Just because people report to the same person or are in the same department doesn't mean they are a team. The dynamic team experience is related to our experience with the co-workers we work with most frequently. A product team, for example, is most often composed of a consistent, cross-functional, cross-skill, or cross-departmental set of people who work together nearly every day on a particular product. A maker team may last for a long time or it may be a short-term project. Irrespective of time, a dynamic team potentially occurs when people come together to make something that matters.

Critical for the experience of a dynamic team are frequent face-to-face, usually in-person, interactions during which members play off of each other's ideas. But often we use the word "team" anytime people gather and coordinate to get something done. People may call themselves a team when they work on a larger project by breaking it

> *The daily life of a dynamic team creates energy by playing off each other's ideas*

into parts, which are then assigned to individuals. Everyone does their part of the work, reviews each other's work, and then someone compiles a deliverable. But this "team" coordination to get a job done rarely produces the dynamic team that women in tech seek. As one developer we interviewed said:

> *When I'm coding alone and just coordinating with others, I still end up feeling isolated.*

So, what characterizes a dynamic, valuing team up to something big? Women on dynamic teams identify with the team's mission, thrive on the daily interdependence of their work, and feel an integral part of moving the work forward together.[46] They psychologically own both their assigned work and the team's overall performance, issues, and success.[16] The Dynamic Valuing Team experience is characterized by in-depth interaction, trust, and valuing each other's contributions. Women are attracted to the dynamism, energy, and feeling of being up to something together, as Joy shares.

> *I worked on the Innovation Team at a large technology enterprise company. It was the most rewarding job of my life. It was thrilling to work on high-level projects with incredible people every day hammering out what we should be doing for next-generation products.*
> *Joy, Director of Consumer Insights*

As Joy expresses, the Dynamic Valuing Team experience creates an atmosphere of creativity that motivates members who value each other to innovate constantly, which is key to their success.[13] Everyone belongs and feels part of the solution. This experience of dynamism, interdependence, and belonging is powerful for team members and managers alike.

Women managers we interviewed also thrive in dynamic teams. Whether they work with the team members or mainly manage the team, they thrive because they too experience themselves as part of that dynamic team. Francine and Nina share their experiences.

Being able to talk to people who are excited about their ideas is what keeps me going every day. Francine, Managing Director

I own the company, but I do the work with my team. I hire the same people for a project that I've worked with before because we all know how to work with each other. I might have to do client management, but the fun is in working with great people to deliver the best product. Nina, Manager of a Consultancy

Managers also thrive because they are in those exciting conversations and interactions that characterize a Dynamic Valuing Team. They see themselves as team members; the one who helps make everything work for the team. They check in with people frequently, attend key team meetings to understand issues blocking the work, are open to questions, and make sure the team has everything they need. These managers feel part of the team's daily interdependence and collaboration. They too are "up to something big" with the team, as Linda shares.

I love my job. I want to be at work even when I'm not. I work with a team full of brilliant people. I listen to their ideas and make sure they can execute. I check in with them continuously, so I know what is going on. Then I can make sure that there are no roadblocks. I feel that I'm a real part of the work even if I'm not designing. I don't do the work, but I make sure they can make their ideas real. Linda, Senior UX manager

Managers can make or break the momentum of the team. Our research shows that good managers encourage risks, accept setbacks in challenges, and commit to the core value that everyone is needed to get the job done. They ensure successful collaboration. They see their role as building relationships between team members. In this way, managers can create the conditions for a dynamic team to form. Interestingly, women managers of large, cross-functional, and geographically dispersed teams seem to excel in creating the kind of team cohesion and communication these complex teams need to succeed.[33, 35] But whether a large complex team or a small team in a start-up, creating and supporting the team can draw managers into the Dynamic Valuing Team experience. And managers want and thrive on this too.

Whether a team member or a manager, the psychological sense of team is what women in tech want and love. The Dynamic Valuing Team experience is at the root of retaining women in tech. And team cohesion is critical for creating it.

TEAM COHESION IN DIVERSE TEAMS

I was in my car talking to my friend before going into work. I was gearing up to walk in the door. I said, "I feel like I have to put on armor before I can face the people I work with. Every day it is an onslaught of challenges to my work, it's like a war." "So why do you stay?" my friend asked. Only then did I realize that I needed to leave, and I did. Jody, Product Strategist

The Dynamic Valuing Team experience is at the center of women's positive experiences at work. Both men and women want that—but too often women don't get it. If women, like Jody, do not feel that they are welcome, that they and their ideas are not invited in and valued, they will leave. If women feel that they have to prove themselves over and over, if their skill and ideas are challenged or simply not sought out, they will not feel part of the team. If women are not offered challenges, if they feel their opinion is ignored, they will be alienated. All of these experiences undermine their sense of belonging to the team, team cohesion, team performance, and retention. Since diverse teams are more creative and productive, ensuring that they are cohesive is a business goal.

Research tells us that without team cohesion and dynamic interactions, diverse teams simply do not develop their full potential and great products are lost.[12, 33] A team is more than the sum of the work of the individuals, the creative interplay between people leads directly to innovation and high performance. When women feel like they belong, they own the team's mission as their own mission. Then the whole team performs better.[33] Team cohesion drives increased involvement and commitment to the team, which in turn increases innovation and performance. Plus, team cohesion increases job satisfaction and reduces turnover.

> *Team cohesion is a prerequisite for high performance and innovation.*

Lower team cohesion, no matter who the team members are and no matter how hard they work, leads directly to lower performance. High cohesion translates into better team performance.[8] When men in a team act as an ingroup, women feel pushed away. As a result, men and women do not communicate and interact in a dynamic way and overall team functioning suffers.[28] Women do not identify with and feel a sense of belonging to their team, so they contribute less to the overall work. In other words, when women are not invited into the team, do not feel a sense of belonging, and are not treated with value, team cohesion and team performance suffer.

Women want to belong to a dynamic team up to something big. But they do not always find this experience in their teams at work. What gets in the way?

INSIDER AND OUTSIDER EXPERIENCE UNDERMINES TEAM COHESION

I was the informal leader of our team figuring out what we should do. So, I thought I'd get the manager job when the lead left, but they brought in Bill from the outside without posting the job. I was really angry and went to my mentor, a director, and told him I was going to quit. He got me a different manager job with a new team and a raise. But then after a re-organization Bill was assigned to work for me! He started sending nasty emails saying he wouldn't do the work. The third-party group hired to help was doing shoddy work and I had to redo it. I still had to supervise the new people and I felt unsupported by my director who said people won't work for female managers. The product was failing. It was a hard decision, but I told

my mentor I wanted to step down and be an individual contributor. This was not fun for me;
I have no interest in being a manager again. Emily, Software Developer

Emily did not feel part of her team. She did not feel valued or supported. Worse, her manager did not believe she was competent because she is a woman. He did nothing to ensure that Bill treated her with respect. Emily was an outsider on her own team. She did not see how she could succeed so she stepped down. When women feel like outsiders, team cohesion is undermined.[10] Unfortunately, identity characteristics, like gender, race, sexual preference, nationality, and age predisposes teams to split into "insiders" and "outsiders".[4, 24] Emily's manager blatantly declared Emily to be an outsider; he was waiting for her to fail.

In tech, women start by default as "not us" because technology is perceived by people as fundamentally male. Research shows that a technologist is expected to be male by both genders.[6,31] The typical technologist is expected to have characteristics and behaviors associated with the male gender role.[7] So, new male members of a team are automatically part of a more valued ingroup. Ingroup members work together more, are more connected to each other, and see the outgroup as "other" and lesser. When people feel that the team is split into "us" and "them" or "not us," team cohesion splinters.[19, 27]

Women and men in tech are perceived and treated differently; they simply do not have the same workplace experience.[22, 30, 37] When a woman joins a team of all men, her gender difference is apparent. Because of our default attitudes about who is good at tech, women often begin a job as part of the outgroup. So, she feels like an outsider. This alienation is even worse for LGBTQ+ women or women of color who are part of two social groups that are stigmatized in technology, making them more inclined to leave the field.[1, 38] Women who feel and are treated like outsiders do not feel welcomed into the team. The lack of welcome undercuts team cohesion from the start.

> *Team cohesion is undercut if women are treated as outsiders*

But team cohesion and connection can also be undercut by differences in interests. Jane naturally shared her male co-workers' interest in games. But this is not always so. When teams are dominated by men, planned social activities meant to promote bonding can instead alienate and annoy women. Back in the 1980s women were told to learn golf and use sports metaphors to fit into the male world. But women today are less willing to make this accommodation. They complain that the teams' social outings are not interesting: sports, gaming, and going out for cigars. Activities meant to build team cohesion end up communicating that women don't really belong. No one is checking in with women about their interests. They are treated as outsiders.

We, the authors, accept that bias is real; that we won't understand another's experience without trying to find out. Unfortunately, the tech culture tips work and social expectations to male interests when women are in the minority. For too long we have expected women to change to fit into a male culture. Leadership development programs have also asked women to change themselves, their speech, style, and other skills to succeed in a male-dominated workplace. But no amount of trying to "fix" women impacts the bias against women inherent in the current technology culture.[5] If we want to retain women, we have to "fix" the team experience so that diverse teams are cohesive and valuing of all team members.

The diverse team is a primary target for the design of interventions. Group cohesion and belonging cannot be depended upon to happen automatically. Without deliberate intervention, women start out in the outgroup. Women will not experience a cohesive valuing team. (See our intervention ideas in Part II.) Feeling in the outgroup undermines team identification, also fundamental to creating the Dynamic Valuing Team experience.

TEAM IDENTIFICATION

At the center of team cohesion is team identification—that feeling of attachment to the people, goals, and work product produced by the team. When all diverse team members identify with the team and its mission, they work harder and bring their varied pools of information, knowledge, and perspectives to the job. So to leverage women's skills, they must first identify with the team. Identification is closely related to feeling uniquely known by team members and the manager. But unfortunately, too often women's capabilities are seen through the lens of their gender. Perceiving a person solely by their identity characteristics, like gender or race, hampers performance and demotivates the team member.[42, 45] Marina shares her experience of getting a new role just because she is a woman.

> *When I was finally asked to take over a challenging project, I was thrilled. I had long felt that my capabilities and achievements had been overlooked and that it was time I was given the responsibility of becoming a project lead. Now my efforts had finally paid off and I felt seen and appreciated. But then I talked to my manager about why I was made the lead for this project. He told me that the project was conflict prone with many difficult stakeholders involved—and he thought as a woman I would be a good person for the project lead. I still can't believe he really said that: We thought you would be good for this because you are a woman! Marina, Software Developer*

Marina was appalled that her sole qualification for the job was being a woman, not her hard work or any of her unique qualifications. She felt that she was not seen as a person; she was simply seen as a member of the category women. She felt her manager assumed she had certain stereotypically feminine attributes he thought were good for the project lead role. Even if Marina is indeed good at mediating interpersonal conflicts, she wants her manager to give her the job because he has seen her demonstrate that required skill. She does not want to be selected because he assumes that she will have that skill because she is a woman. Being attributed these qualities based on gender made Marina feel invisible. It undermined her sense of belonging and identification with the team.

When managers and team members make decisions based on a person's social categorization instead of their unique qualities, team cohesion and team identification suffer. On the other hand, research shows that a strong team identity is built when team members feel known as individuals and are given work based on their unique qualifications.[20] Being known for our uniqueness also means taking the time to know us. This is a form of the feeling of being valued that we discuss in Chapter 10.

To build team cohesiveness we must also check ourselves to keep from making decisions based solely on our assumptions related to an identity category. Cedric, a development manager, shared his embarrassment when selecting a buddy for a female Asian-American new hire.

> When I'm working with new hires, I think about what helped me when I was new. I'm French and bonded with another young person on my team because we both spoke French. It helped me make a friend and learn about the team. So, with my Asian new hire, I paired her with a Chinese guy. I'm embarrassed to say that it was a disaster. She wasn't from China. They did not speak the same language! I wanted to give her a friend by making sure they had something in common. But they didn't click at all! She wanted someone responsive to questions and he wasn't at all. *Cedric, Development Manager*

Cedric had good intentions. But it backfired because he treated his new hire as a member of the category Asian. He did not know her as a unique individual. His new hire wanted someone responsive, but she was paired with someone who didn't talk or make himself available to help. Without seeing both of his employees as individuals, he categorized them and failed to plan a real connection.

Unfortunately, our propensity to see people as category types is further reinforced by remote working. In a remote session, communication is more prone to stereotyping because of the reduction of cues.[25] When people communicate through remote technology, they do not hear as much nuance in another person's voice. They miss non-verbal communication and body language. They can't see movement or another's attentiveness. This is particularly exacerbated when people meet with video turned off, as we have found many teams do. Distributed teams who used to meet remotely but in small, co-located groups without video tend to continue that practice.

If individuals are seen as categories—not for their unique skills—they won't feel part of the team

But since people are social beings, when remote they grasp any cues they can get about the other person, trying to form a coherent impression.[47] As a result, cues about group membership—men, women, older people, race—become more important in online communication. Identity characteristics like these play an outsized role in forming an impression of the other person when remote, further undermining getting to know people in their uniqueness.

But focusing team members on team identity can disrupt thinking of others as a category type. Team identification and commitment to the mission can build the common ground necessary to allow each team member to emerge and be seen as an individual. Interestingly, visualizing something that represents the team, for example through a team avatar, can help members focus on team membership.[40] Being known for our uniqueness fosters a sense of belonging that helps create team cohesiveness. A diverse team with a strong team identity encourages sharing and using the different knowledge and perspectives of all team members. Then these multiple perspectives are more likely to be discussed and integrated in the team's tasks and goals[43] resulting in increased team creativity and performance.

TEAM COHESION AND REMOTE WORK

Remote work is not new to the technology industry, but 100% remote work is.[36] During the pandemic, each person was often isolated from all co-workers interacting only through collaborative tools, sometimes with video and sometimes not. Research on remote teams conducted before the pandemic found that remote work undermines team cohesion.[26] When all team members are remote, the conditions for creating dynamic teams are even more challenged. As an industry, we are engaging in a living experiment as some companies choose to remain all remote, some will be primarily co-located again, and many will be hybrid.[36] To retain women in tech, to create valuing dynamic teams, we need to understand the impact of remote working on teams.

Findings from our Remote Work Project[3] suggest that team members who were co-located before the pandemic and who already had good working relationships, can more easily transfer those relationships to the remote context. But new hires, or people working with that new person, consistently report how much more difficult it is to build new relationships in a remote context. As we said, remote working challenges the creation of social connection. But it also challenges critical informal interactions. One-on-one drop-in interactions for chit-chat and work questions, spontaneous brainstorming, and problem solving have been cut off. Group interaction and more structured working meetings are the primary contexts for collaboration. Informal conversations that occurred before and after meetings have also evaporated along with hallway conversations making collaboration inefficient. Ginny, based in the United States, bemoans the lack of collaboration with and between her development team in India.

> *I have a team of three developers in India who used to sit together but are now each working remotely from their home. Before, the developers each took on stories to develop but in reality, they worked on them as a team. If someone got stuck, they traded around stories, all working on the set together. If they had a quick question, one of them picked up the phone or sent me an*

---

3   Finding related to retaining women in tech are summarized in the Conclusion. Also see an overview video of initial findings from this project https://www.witops.org/the-remote-work-project/.

instant message. It was very efficient. Now they don't collaborate at all and they don't check with me on little things. The guy who used to call sometimes still does but not the others. I don't think they were that comfortable talking before—he was the designated caller! Now there are more bugs. If someone has a question or gets stuck, they write a ticket; we have so many tickets that everything is slowed down. It is very frustrating. We are trying to encourage them to reach out to each other informally but that means asking for a meeting or, I think, admitting they aren't sure what to do. Ginny, Technical Product Manager and Product Owner

Ginny tries to encourage informal interaction but is at a loss of how to get the collaboration back. She also reports that sometimes people who used to collaborate to get to an agreed design before a meeting, now take the meetings off track because they end up collaborating in the meeting while everyone else waits. Simply being co-located fosters collaboration and a more dynamic way of addressing work problems. Our research on remote work revealed this overall loss of informal collaboration and that co-workers mainly communicate about what they are doing in formal meetings, or as in Ginny's case the ticket tracking system. Other research on co-workers that were previously co-located and are now working from home also finds that remote working negatively impacts team performance.[41]

Remote working undermines informal communication and connection

And the lack of informal collaboration, especially dropping in, can cause friction. One content architect was furious because she didn't find out that the video specialist was developing a video until she was in a status meeting. Previously, she was consulted about any upcoming video work. Then she could provide input to ensure that the video was developed to web standards. But they coordinated when the video specialist dropped into her office to "chat" which included work check-ins. Since this specialist doesn't like to use instant messaging, he doesn't "drop in" now. Drop-in interactions that often contain both social interaction and work clarifications are gone. When remote, this easy way to communicate is lost.

Remote working makes creating a cohesive connected team much harder. In both of the above stories, everyone knew each other and had previously worked together. But in the remote context, their ability to collaborate is still undermined. So, what happens with new hires? New hires need to build connections and learn the styles of the people they are going to work with. They have to learn the job and collaborate with new people to come up to speed. If everything and everyone is new, and if face-to-face and informal interactions are not possible, team cohesion is difficult to create. Lein, an operations manager based in Manila, tells us about her challenge when starting a new role on a global team.

I'm working with leaders in three parts of the world and I have never met them or their teams. Before the pandemic, we traveled to each location several times a year and spent time face-to-face working and socializing for a few days. We got to know each other as people and our work styles, which helped in our remote interactions. Now, especially with our practice of having cameras turned off, no one is real. I don't know what they look like. It's harder to make my presence felt and read their reactions. It's my job to make sure these offshore teams work well, having no face-to-face interaction makes it difficult. I have to exert extra effort to try to read them and help the team collaborate—I just want to jump through the screen to connect. Lein, Senior Operations Manager

Lein had worked on remote teams in her previous position. Before the pandemic, this global company built in ways for people on remote teams to become "real" to each other. They met multiple times a year to connect socially and collaborate. Frequent face-to-face interactions facilitated the success of remote teams, which helped overcome some of the known issues with remote teams.[32] But now, without any interaction with new co-workers, Lein feels ineffective and disconnected from the people she is managing.

Remote work can increase stereotyping by reducing the interpersonal cues necessary to connect and perceive each other's unique contributions. It all but eliminates drop-in communication and discourages 1-1 communication because a meeting must be set up. New hires find it harder to connect and to get help and coaching. New research also suggests that women working remotely in the pandemic continued to report that they did not feel heard.[14] Add to this the consistent finding that women were also taking on a greater burden for kids and home life during the pandemic.[11] None of the conditions found in remote work during the pandemic help create a more cohesive team. To retain women in tech and get the most out of our teams we need to create cohesive dynamic teams. As an industry, we need to monitor and experiment with ways of working that create and grow team cohesion. (See tips to foster team cohesion.)

NUMBERS MATTER

Dynamic teams create a sense of belonging and value for all members. But as we said, this does not necessarily come automatically to women. The default male culture created in tech companies naturally occurs because most of the members of the team are men. Individual women on teams can feel alienated simply because they are alone. Individual women on teams run the risk of being seen as tokens,[21] especially in the tech culture.[2] The lone woman serves as a visible representation that women are actually in an organization, but they are often not known as a person. They are too often seen as a representative of all women. And because their difference is highly visible, they can be marginalized for being different. As tokens, lone women are scrutinized and often doubted or not trusted.[39] As a result, they may be uncomfortable, isolated, or have self-doubt.[21] Not surprisingly, the feeling of being an outsider can interfere with their performance[34] and, of course, undermine team cohesion.

Foster Cohesion on Remote Teams

We are all experimenting with how to manage remote diverse teams. Here are some tips based on our experience and research.

Encourage social connection. Keep up the group coffees and personal chit-chat at the beginning of meeting. Encourage everyone, and especially new people, to share fun facts about themselves including hobbies and other interests. Use icebreakers and warm-up games before workshops. Beware of gendered social group experiences—assign a diverse team to plan it. And don't stop these social interactions with remote people after returning to in-person work.

Foster 1-1 connections deliberately. People connect through working directly together not just in a group. Assign work to teams and rotate who works with who. Senior-junior and junior-junior partnerships are great for new hires and help people get to know each other more. To foster cohesion when remote, think beyond what is necessary to what will increase communication and skill. Beware if you ask for volunteers, people will want to work with their friends. Foster relationships by assigning people who don't know each other well to work or plan together. Make it clear that team cohesion is a business goal.

Support team Identification. Since we lose cues when remote, we suggest keeping the cameras on. Get a team commitment on this and enforce it. Focus on knowing and making decisions based on the real person, rather than their categories. Highlight and celebrate unique contributions and watch out for gender-based assignment. Display symbols of team identification and belonging, e.g., have a team video background, develop a team logo and use it on your online whiteboard, and create avatars for each team member as a variation of a team mascot. Try creating and displaying a Team Manifesto to remind people of the kind of team they want to be, include a commitment to practices that disrupt bias. (See Chapter 11.)

Encourage and model informal drop-in communication. Tools like instant messaging are potentially great for dropping in but not if, as our data suggests, people, especially junior people, feel that they are bothering others. Try turning off "do not disturb" and allow interruptions or keeping the door open more often. Encourage sanctioned "let's have a coffee break" messages to encourage 2–3 people taking a break together. Be creative to recover drop-in connections and information exchange.

Get together face-to-face. Fully remote undermines the dynamic team experience. Ensure that all members of the team get together as a group and as pairs often enough to work and chat together. Make sure no one is left out.

This outsider experience can change women's perception of daily interactions. We can start looking to see if we belong. The first meetings with the team and each interaction are examined to see if we will really be welcomed and valued by our team. Being welcomed into the team is one of the key ingredients we found necessary to help any new hire connect and succeed. The Welcome, as we call it in Chapter 7, can start to sweep this away.

> *I am the only woman of color in the department I have just joined, I was nervous on the first day. But a senior female colleague swept me right into their world. She was so excited to see me, to have me come to their department. She invited me to meetings, went with me to events, and generally made me feel like I belonged. She opened the door to their world.*
> *Ivy, Assistant Professor in a technology department*

When someone from the ingroup opens the door and validates our worth, we start to believe in our own belonging. When there are several women already on the team or in our group the numbers alone communicate The Welcome. The women we interviewed continuously report the importance of having more women to work with. Simply seeing like-me people on the team, in management, on the corporate board, on the walls of esteemed organizations sends a message of welcome. Seeing others like me shows underrepresented people that they are invited to participate in this team and company.

More women on the team changes everyone's experience. With more women, everyone must go beyond their gender category when thinking about how each woman may participate. Each woman is more likely to feel known as an individual. Often with more women, social connection improves. With more women on the team, the culture starts to shift away from the "boys' club" women so often point to as a problem in tech culture. So, more women on the team can improve the conditions for creating the Dynamic Valuing Team experience.

Our research also shows that a more balanced gender mix improves the everyday work experience for all members of the team—both women and men. Figure 1.1 shows the results from our survey. With more women, everyone on the team experiences an improved work experience. [18, p.153] Everyone on the team thrives. A more gender-balanced team has the potential of nudging teams into becoming more dynamic and cohesive and therefore more innovative and productive.

Our finding is in line with the vast amount of research investigating tokenism.[17, 21, 48] Once women make up about one-third of the team, the token experience vanishes. With more women, social norms change for the whole team or the whole organization. And once the number of women crosses a critical threshold near to 50%, the productivity of the team increases.[3, 29] Given a reasonable team size, a minimum of three women is seen as the "magic" number where women's experience normalizes and hypervisibility is reduced.[23] A single woman on the team might make the team technically diverse but it may not reap the benefits of a diverse team. As our and other

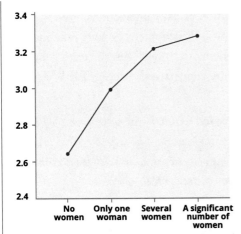

Figure 1.1: Scores for the scale "Overall Experience" based on number of women in team.

research suggests, the overall work-life experience and productivity of diverse teams improves with more than one woman on the team.

Team cohesion, performance, and culture all change when more women are on the team. Short of instantly hiring more women, a simple intervention is to change how we distribute women onto teams to ensure that women have the critical mass necessary to change the team culture. What might that look like? Instead of focusing only on the company's overall numbers, focus on the configuration of the working teams. Require any working team to have a maximum of 50% of the current majority group—in tech typically white men. Make sure there are never fewer than three women on a team. Start with the people you have now to set expectations each year. Then increase the expectation each year and set a date by which no all-male teams will be allowed. This kind of policy can also push hiring. "Culture fit" criteria are too often a way to hide gender and racial bias leading to more "like me" hires.[9]

We know that if we hire more women into technology that numbers alone will help shift the culture. And we know that having more women invites more women to take the job. Stories from the women we interviewed clearly indicate that they look for the number of women in management, in their group, and on their team when taking a job. Women want more women to connect with and as an indication of the company's commitment to diversity. On the other hand, we have been trying to hire more women for over 40 years. We have to learn how to change our practices to create dynamic valuing team no matter the number of women on a team.

> *A more balanced gender mix—not one woman on a team—improves the work experience for the whole team*

THE PRIORITY OF CREATING DYNAMIC VALUING TEAMS

The quality of our interactions with co-workers makes or breaks our experience at work. If each person feels valued and embraced as an integral part of the team's success, they thrive and the team performs. Ensuring that every team member experiences belonging and value no matter their identity characteristic is the only way to reap the benefits of diverse teams. And when all the team members are positive about the value of diversity, it has a dramatic and positive impact on team performance.[44]

Getting more women into tech and onto every team will help. But we must accept that women will not automatically feel welcomed or have a sense of belonging if we do not change team practices. Instead, we need to build teams and organizations that make The Welcome and the experience of the Dynamic Valuing Team happen on purpose. As we said in the Introduction, successful teams have clear goals, roles, processes, procedures, and agreed standards of goodness. Throughout this book we will introduce a variety of intervention techniques that will help foster tight-knit dynamic teams that get great work done. Good management is the ability to create the Dynamic, Valuing Team by design. This is a big step toward retention.

The experience of belonging to a Dynamic, Valuing Team is central to what helps women thrive. Its importance for retention threads through all the factors in the @Work Experience Framework. Dynamism is about how teams interact to get the work done. But they also must be up to a team mission that matters to everyone. In Chapter 2, we address the second factor: the need for stimulating work.

CHAPTER 2

Stimulating Work

I joined a company that was on the cutting edge of engineering dealing with the newest tech for communication protocols on the Web. The architects worked on tough problems. I was one of very few women on that team. It was a culture of competition, yelling, and challenging each other's ideas but it wasn't personal; even the smartest guy got yelled at. Stuff got done and we really collaborated. I want to be where the decisions are made, where you can influence the state-of-the-art tech and there are so many smart people. Jen, Software Development Manager

Developers, designers, project leads, and user researchers—all the women we interviewed—want to work on real tangible products; projects that are important to the company, the industry, or the world. Women want high-impact and high-profile challenges. They do not want to be bored and switch jobs if they stop learning and feeling challenged. Women in tech leave when the assigned work does not match their skills and career goals. If they feel unsuccessful, ill-used, or under-challenged, they leave. Women value the Dynamic Valuing Team experience, but they equally value stimulating, challenging, and ever-learning work.

When Jen took the job at this leading-edge company, she knew what she was getting into. The culture was challenging, the women were few, but she wanted to be where decisions were made about an emergent technology. The problems were hard, but she got to work with the smartest people. Jen first became a manager in that company, managing those very smart people in this demanding culture.

Jen's experience, like all the women we interviewed, stands in contradistinction to stereotypes of women. It is often assumed that girls place higher importance on social relevance and that this shapes their decision to later enter STEM. But when the impact of social relevance on students' career choices is actually measured, the results show otherwise. For computer science majors, there is no difference in the importance of social relevance between girls and boys. Differences do exist for choosing STEM subjects like biology or physics where social relevance matters for girls more than for boys. But in computer

Figure 2.1: Results from our first survey, confirmed in our second survey, showing the reasons women like tech work. Social impact is lowest.

science, perception of social relevance has the same impact on choice for both genders.[1] These findings are in line with both our qualitative and quantitative data. Women are interested in the most difficult technology; they do not need socially relevant work to thrive. Women thrive when they have the same challenging work that men seek.

Women who work in tech do so because they have a high interest in just that, tech.[11] Figure 2.1 shows the results of our survey for the @Work Experience Framework revealing the lower position of social relevance. Women describe their work as multi-faceted providing them with opportunities to learn, be stimulated, make an impact on the company, create something innovative, and work with great people. Managers see their work as meaningful when they help create productive teams that achieve their mission. Neither managers nor team members who report a positive experience at work see getting or doing socially relevant work as important to their own well-being. Socially relevant work may be one source of meaning for some people, but it is not the primary or only driver of women's interest in their work.

The Stimulating Work factor highlights the role of work challenges for retention. Like men, women need the opportunity to do highly engaging tech work. They do not need socially relevant or special projects tailored to attract women. Like Jen, women developers love bleeding edge technology and hard problems to untangle. Jane shares her love of coding.

> *I love every kind of puzzle and coding is just a big puzzle and that is fun. I'm so completely absorbed in finding a bug or the technology challenge that nothing else exists in my mind. It can feel brutal, but then when you solve the problem, it feels great, it's a high. Jane, Software Developer*

To Jane the puzzle is totally absorbing. Solving a problem can be frustrating but when she is successful, it is a high. The high gets her through the problems. Her ability to solve the problem builds and reinforces her feeling of competency and reveals that competency to co-workers. But solving a problem is also interesting because it is also producing something tangible. Francine who works in the automotive industry shares her love of making something that people will use.

> *From the beginning of my career, I loved working on cars because if we are successful, the cars will be cooler than they were before. And everyone has an opinion about car design. It feels very relevant to everyone. Francine, Managing Director*

Shipping a product is putting something new into the world that will affect people's lives. For Francine, shipping a product like a car is shipping cool technology and making something relevant to everyone. The tangible nature of a product motivates many makers. They want to make something that is substantive, that they can point to and say I helped make that. But making isn't just about coding or physically building a product. For Beth, making is about the organizational challenge of getting the product out the door.

The fun part of my job is the implementation and rollout of the product because it becomes real. Every new job I've taken has been with companies making a new transition, for example from print-based to technology products and then from one technology platform to another. I thought about getting a job in a think tank but that doesn't produce a real product that will ship. Helping an organization ship a product gives me a sense of pride. Beth, Director of Product Innovation

Beth doesn't want to think about what to make in some future life. She is challenged by all the organization and decision-making that goes into pulling multiple teams together to actually ship something. Beth is attracted to organizational and motivational challenges critical to bring a product into being. Women in tech need challenging problems appropriate to their job roles. Eve, a user researcher, is also excited by her role in making something real.

When I was looking for my new job, I knew that I wanted to be part of an initiative that was up and running; something concrete that was going to ship. I was done being part of advertising; building a brand isn't building anything real. Eve, Principal User Researcher

Eve had been in an advertising firm, producing ideas for a brand. When looking for a new job she wanted to create something concrete that ships. User researchers see themselves as making or improving a product. Understanding the customer, finding latent needs and ways to improve the product, and working with designers and developers as a cross-functional team are all parts of technology making. Technology jobs span many kinds of challenges all of which are needed to create a successful product. Whatever their job type, women thrive when they are part of a dynamic team creating something that matters for the customer and the business.

> To retain women, give them highly engaging work

Making products by themselves is a great draw but Beth's experience also reveals the draw of business challenges. Beth loves helping a company transition from a print to an online product or from an old platform to the latest one. Robin is also attracted by solving business problems.

I went to a large well-known internet company because it was failing. I was attracted to the challenge of generating ideas that would save it. We had to get the right balance of what would work for this business and compelling technology which would lead a new direction for the company. Robin, Engineering Manager

Robin was excited by the challenge of helping to rejuvenate a failing company. She was tasked with generating product ideas that could both attract new customers and produce revenue. But she was not alone in this task, she was part of a dynamic team. Together they generated innovative ideas through a dynamic interplay fueled by her manager's push toward out-of-the box thinking—just the kind of challenge she thrived on. Building the business is also a technology challenge.

Whether solving hard problems, getting the right research to improve the product, figuring out what to make to support both customer satisfaction and business goals, making a company more successful, or impacting the industry, women need work that is stimulating to them. Technology as a field is a draw because it is ripe with technical, customer, design, and organizational challenges for women to solve. When women are in jobs that provide these challenges, women thrive. But what happens when they get bored?

BOREDOM AND RETENTION

> *After nine years working on one product, I wanted out. My boss said, "Don't go, I need a good QA person." But it was enough. I decided to work for a graphics company because it was different, but also, because I used two of their graphic products and respected the company. Ellen, Quality Engineering Lead*

Ellen was bored—no other complaints. She went to the graphics company for something new and stimulating—and because she knew and liked the product. The new company was both a challenge and personally meaningful to her. Our survey shows that people thinking of leaving their jobs report greater boredom with their work. If women are not intellectually engaged, they will leave. Unfortunately, according to employee engagement research in 19 countries, only 16% of the employees are fully engaged at work. An astonishing 84% are just going through the motions.[3] Even with benefits or other perks, even if they are connected to their co-workers, boredom drives women away. Beth shares her motivations for leaving one of the companies she worked at.

> *I left the company that paid for my MBA within three months of getting the degree because the job was a dead end. I didn't see where they were going, and I wasn't learning any more. I want to be in a job that exposes me to new ideas, new technologies, and new smart people. Beth. Director of Product Innovation*

Beth wasn't learning any more. She stayed to finish here MBA and then started looking around. Beth's experience of a dead-end job was not related to promotion; she was looking for more stimulation from the work and from new people. Beth had worked at this publishing company for a while and saw the emergence of technology to support publications. She did not see this company moving toward new technology platforms fast enough, a key attraction for her. So, she left and took a job to help another company move to a new technical platform. Joy also leaves her job simply because of boredom.

> *When I left my UX job I went into marketing designing a viral marketing campaign. There's no way I could do that in UX. I was looking for really varied work and this was exciting and new—until I got bored with that! Joy, Director of Consumer Insights*

Joy makes it clear that boredom is inevitable. No team experience can offset her need for stimulation. Joy will not stay in a job if the work is not varied, exciting, and new. If companies want to keep valued employees, and especially women, they need to pay attention to whether they are getting challenging and stimulating work. Companies cannot wait until women get bored to intervene—by then they are psychologically out the door. Unfortunately, the pandemic may not have helped the situation. Amy shares her experience.

> *I work with the branding teams, but there is a lot of in-fighting right now, so they are not putting forward new projects. Many brands are holding back on creative work. One of the most frustrating things is waiting to be chosen to work on a new, exciting project. Amy, Senior Digital Designer*

Amy is waiting for the business to regroup and feel stable enough to start new projects. She is understanding about the challenges of work during the COVID pandemic. But Amy is also thinking about her future work and job. Amy has been advocating to be the new art director and feels supported by her manager. But everything slowed down during the pandemic. Everyone we interviewed in The Remote Work Project expressed tolerance and appreciation of the difficulties faced by companies and co-workers because of the restrictions during the pandemic. But still, women at this time want and need challenges and get bored. If not, they may leave. Gail took a new job during the pandemic. She shares how her previous company lost her.

> *I was at a satellite office in San Francisco after my last rotation out of college. But the bulk of the developers were in Virginia. We were always being second-guessed; our work was redone or just thrown out by the home office team. I had no feeling of making anything real. Gail, Software Developer*

Gail left because she wanted to make something tangible—but nothing she coded produced a real product. Coding for no purpose, getting paid but not making anything, wasn't enough. By failing to ship the work from her satellite office the developers in the home office sent the message that Gail's work was not valued. Even if the problems she solved were engaging, Gail wanted to make something real that shipped.

Boredom is directly related to retention. Being bored is one of the fastest ways to drive women out of a job and onto something else. Stimulating work, work that matters, work that produces tangible results, work that keeps women learning is required for women to thrive. But it is not just required for individual women, it is also required for the diverse team as a whole.

CHALLENGE THE TEAM AND THE INDIVIDUAL

> *I'd been working at my company as part of a great team. But when the company was acquired, they outsourced the challenging work like making a mobile app. One of the guys on my team left and started recruiting the rest of us to join him. They finally recruited me, and I negotiated the kind of work I wanted. I was just waiting for an invitation to follow my team. Robin, Engineering Manager*

Robin worked on a new technology for many years. She fought to get the job because it was the leading-edge product of its type at the time. She helped get the first release of the product up and running at her first company. She became a leader in that product community. After many years, the work had become boring. Her team was dynamic, very cohesive, and had been doing interesting work continuously. But now Robin and the whole team was bored and started to leave.

When the whole team is not challenged, the whole team can gravitate away. It's not surprising that in Robin's situation, the stimulus for leaving is an acquisition. Instead of offering the next challenging work project to the existing high-functioning team, the new company managers gave

Manage the stimulation of the individual and the team for retention

it to a third-party consultancy. By not giving this or any interesting work to the team, the company implicitly sent the message that they did not value the productivity and creativity of this team. So, they all left.

As in Robin's case, the first person to leave a dynamic tight-knit team often recruits their favored team members. Gail was recruited away from a dissatisfying job by people she had worked with. When a manager and several teammates went to a start-up, Gail joined them. Janet also built her new team by recruiting members of the team she had built at her previous company. She tells her story.

> *My VP was increasingly unsupportive. She had a graphic design background and had a hard time understanding how to leverage UX skills. Worse, she didn't understand the Agile process that we were now using for development. If they asked her for five new features, that is all she wanted from our customer research. She did not understand user research and started cutting our budget and undermining me in meetings with senior people. Eventually I decided that I had enough—I was working enormous hours and getting no support. So, I left, joined another company, and started hiring my best people over to the new company.*
> *Janet, Director of User Research*

Janet felt that she and her team were not valued; that they could not do their work with quality—so she left. She had built up a team of talented people. When she settled in her next job, she started recruiting her old team to reconstruct that team experience elsewhere. From Janet's perspective, the work of the research group was undermined by her manager. As a director, stimulating work for Janet is building teams. Without support of her VP, Janet left and pulled the team with her.

We all know these stories. One person or a manager leaves a company and then recruits their favorite co-workers. Transplanting a dynamic team into a new challenge elsewhere ensures that the people will get up and running fast. The cohesive, dynamic team is recreated but now focuses on new challenges. This is a classic recruiting technique, hire one good person and then encourage them to reach out to their known network to fill available jobs. If companies do not want to lose women in tech, if they do not want to lose the dynamism of a creative team, managers must manage the stimulation of the individual and the team.

Managing the stimulation of a team as a whole is very different than managing individual careers. This is especially urgent for cohesive, dynamic teams who need to be continuously challenged as a team. Think of a dynamic team as a unit; manage them and their work as a team, not a set of individuals. Plan to give them another challenging project to benefit from their innovation and productivity. Without compelling work and engagement for the whole team, the team loses focus and the interest in the job which is needed to drive the work forward. If the team gets bored, the people will start looking for something new elsewhere.

Research on teams help us understand how to build team commitment to a new project. Managers can instill purpose by making sure each person on the team sees the value of what they are doing both individually and collectively. A commitment to the mission requires feeling a part of something more meaningful than just finishing a user story, doing the required AB testing, or meeting a deadline. A dynamic team becomes engaged when everyone has a shared answer to the "why" of the project; how the work is relevant, what they are contributing

> *If teams are bored, they lose commitment to the team mission*

to its outcome, and the impact of the project.[25] Once a cohesive team is engaged, their collective sense of purpose creates the energy to drive them forward and spark creative collaboration, which in turn reinforces that sense of shared purpose.[6] Engaging a dynamic team on a new project starts with getting all members of the team bought into the value of their project.

But if women on a diverse team do not feel valued, they will not develop a sense of belonging. This as we said, undermines cohesion. The team will not develop into a dynamic team because women will not be fully engaged. Valuing the contribution of each individual on the team is central to creating and maintaining a dynamic team. And value is communicated by the type of work women are assigned. The Stimulating Work factor is central to their retention. But if women do not get challenging work because of bias, both retention and the ability to create a dynamic team is undermined.

WORK AND GENDER STEREOTYPES

Women need challenging, stimulating work—just like men. Research shows that men and women do not differ in their willingness to perform challenging tasks.[8] Unfortunately, both women and men still hold stereotyped assumptions about what women want and are good at.[9] These stereotypes get in the way of women getting the work they need to keep them engaged.

Because women are expected to be more social and work better with people, women in tech are often steered into non-technical jobs.[26] Women may end up in marketing, people management, or coaching roles instead of more respected technical jobs. User experience, design, user research—the whole field of human-computer interaction—may be seen as good for women because it involves interaction with and understanding of people. Beneath all these notions of what women will want to do, what they will be good at, is the assumption that any woman is a people-person— even a highly technical introverted woman.

The social definitions of what women are good at shapes the work they may get and take—and whether they will choose to be in a technology field.[14] Our stereotypes of tech work communicate both to women and their co-workers that women's skills aren't good enough for technically hard jobs.[4] Years of research have established that boys are not better than girls at math and technical things, but only that we encourage each gender in different directions. Initiatives like STEM schools, camps, and organizations like Black Girls Code that encourage more girls to go into STEM[22] deliberately attempt to disrupt this stereotype. They deliberately change the training and expectations about what are appropriate interests for women.

> *Don't assume that women in tech are good at or want work with a social dimension*

If we change education and cultural expectations, women's choices do change. For example, in India computing is seen as a women-friendly profession that offers a professional career and a safe and pleasant indoor working environment.[27, 29] Women coming to the United States from India for education are more likely to feel that technology is a valued profession and that they are technically competent.[28] Similarly, when parents decide to support their daughter's STEM career-related learning, a girl's expectations for herself can change.[16] Nearly every woman we interviewed told us a story of how a parent, relative, sibling, or teacher was both the role model and the champion of her interest in all things technical.

But societal expectations are hard to change. Women internalize messages about their own technical competency, creating a self-fulling prophecy. As a result, they may hesitate to take on hard technical tasks. Once on the job, women (and men) bring their stereotyped self-expectations and feel pressure to conform to them. For example, women may avoid being assertive. They may feel responsible for childcare or gravitate to certain tasks like planning team parties.[10] Even when women choose a technology job, get a Computer Science (CS) degree, and feel competent, they may be working with managers and co-workers whose unconscious expectations of women's technical competence have not changed. This bias works against women getting the hard technical tasks. If, as a result, women feel devalued as well as bored, they will leave.

> *Societal expectations send women the message that they are not technically competent—but they are*

COMPETENCY AND STIMULATING WORK

> *When the start-up got bigger, more engineers came on board, and they were men. I felt they were more comfortable with each other and banded together. They resisted me in every daily decision. One man who reported to me said, "I'm an engineer and you are not."*
> *Beth, Director of Product Innovation*

Beth does not have a CS degree and so was treated as less competent to make decisions. The male developers did not recognize her legitimate authority to lead them even though that was her job. Like many of the women we interviewed were aware that their CS degree and tech skill gives them credibility. Janet was also clear that having technical authority is a key to success. She contrasts her own effectiveness with her manager's ineffectiveness.

> *My VP is a designer and does not understand how to talk to developers about the work of user research and design. Without a technical background she has no credibility to engineering and product management. She speaks in a "flowery" way instead of the direct, no nonsense, fact-based style required to get through to tech people. Janet, Director of User Research*

Both men and women perceive their credibility through a technical capability lens. Those who "own the code" automatically have power and prestige; those that don't have less influence and control.[5] Janet knows that her way of communicating helps her demonstrate the value of her team's work to her technical partners. She knows that the ability to "talk tech" gives her immediate credibility and underlies her influence. Her manager does not have these skills or credibility to effectively advocate for UX work. In Janet's opinion, her manager shows up like a stereotypic woman.

Women are expected to be interested in social matters while men are expected to have technical competence.[7, 13] Janet implies that to be successful women need to be direct, no nonsense, and fact-based—all styles that are typically associated with men. Women have long tried to build their credibility and belonging by changing how they speak and behave to be more masculine.[4] This equation is at the core of the myth of meritocracy in tech: Masculine behavior is taken as a sign of technical competence— not actual skill. Competence is not really about skill, but about behaving a certain way.[24] But as long as this myth persists, women have to play along. If women by default are not seen as technically competent, they may not get stimulating technical assignments.

Masculine behavior is seen as a sign of technical competence—it's not actual skill

So, when managers in tech are faced with a decision of who to assign to a hard problem, they default to a man. Managers might assign the work to a man with whose work they are familiar, often one with more experience. Even when faced with two new hires of different genders, managers are more likely to pick the man for a technical issue. They assume the male will be the more competent even though neither new hire has yet demonstrated their skill. Women are assigned more nonchallenging tasks.[8, 19] And women may have their competency questioned, as Anne shares.

> *My male colleagues are constantly questioning my work. I do not think they would have treated a guy like that. So, I try to act like one of the guys. I wanted to fit in. If you are different, you do not belong. Anne, Software Engineer*

As a woman, Anne's work as a software engineer is questioned again and again. She has the requisite CS degree but that is not enough to give her the default credibility she sees afforded to male colleagues. Women feel that they have to prove themselves over and over—that they are starting from scratch with each new person.[20, 32] Constantly proving herself undermines Anne's feeling of acceptance and belonging on her team. So, Anne tries to act like one of the guys, not her normal style. Over time, this pretense is draining. This self-alienation, changing one's behavior to conform, is unique to women, people of color, and LGBTQ+ employees.[4, 23] Anne changes her behavior because she is not perceived as technically competent by her team.

Women's competence is questioned again and again—it's bias

Without an intervention, men can get the better work by default. Because men get good work challenges, their work is more likely to be showcased. Perhaps this contributes to the finding that women are less likely to be promoted into higher technical roles.[26] Our stereotyped expectations impact task assignment, which impacts women's career development, which further impacts retention.

These issues of task assignment are exacerbated by remote work. As we said, working remote reduces the stimuli we receive about co-workers which encourages people to fall back on stereotypes like gender.[30] When remote, people tend to forget what they know about each individual woman's contributions. Instead, women are assigned to tasks based on assumptions of what women are good at. So gender influences whether women are seen as suitable for an assignment.[12] Unfortunately, being remote does not render gender invisible; it does not encourage recognition of a person's unique potential as some had hoped.[15]

Whether remote or in-person, bias affects the tasks that women get. For any given task, the most competent person may not be assigned to do it. When women feel that their competency is questioned either explicitly or by not getting challenging work, they feel devalued. The tasks women get affect their sense of belonging, team cohesion, and so engagement in the team mission. To build dynamic teams and retain women in tech we need to distribute challenging work more equally. (See tips for distributing work.)

Distribute Stimulating Work Equally

Tips for remote or in-person teams

Track who is getting what work to see how gendered assignments are. If the men get the hard jobs, if you trust them more, examine your bias.

Round Robin to change your behavior. Pass the hard jobs around equally. You won't know if women or less experienced people excel if you don't try them. But be sure they have a go-to senior person for help.

Check who is getting promotions and who are not. Ask if it is because they are not assigned the hard problems. Maybe they are working hard but have difficulty asking for challenges. Plan to give them more challenges.

TEAM HOUSEWORK

Both the start-ups I worked in and the research facility were mainly men. The guys there asked me to do gender stereotyped stuff like the party planning or make cookies. I just refused. Eve, Principal User Researcher

In my experience, women are more willing to do the unsexy grunt work, and they are there-fore given more of it. Whereas males are more likely to say, "I don't have time for that." Meg, Senior Engineer

Every team has a set of activities that keep the team going: taking minutes, bringing food, party planning, making coffee, grunt work of one kind or another. Stereotyping fuels expectations that women will welcome doing this work—and many do. Eve had her strategy—just refuse to do it—which Meg says is typical of men. Other women enjoy being the "mom" of the team, bringing the cookies and planning the parties. Whether women like to do it or not, social housework takes up women's time that would otherwise be used in career enhancing project work.[2]

The perception that women will like team housework can extend to requests to help with hiring and mentoring. Eve shares her resentment.

I was in a small start-up with few women. I am currently mentoring two women, one is my research assistant I brought from my old company, and the other intern was assigned to me to be mentored. I feel like they just assigned her because we were both women. Then they asked me to hire a female intern in order to create diversity but the men were not expected to do this. Just because I'm a woman doesn't mean I want to hire and mentor all the women in this company. I don't have time for this. Eve, Principal User Researcher

Eve does not want to be the default diversity manager expected to hire and mentor women. Eve resents the expectation that she will do this just because she is a woman. Hiring and mentoring takes up time, taking Eve away from doing the actual work. Asking women to mentor or hire other women treats them as a category type, presuming they will want to take on diversity work which does not directly lead to career success. Plus, to build a diverse workforce men must be involved in hiring and mentoring. If men have more influence in the company, women need male mentors to enhance their careers.[18] Mentoring is important for career development, but women do not have special mentor skill. Instead, companies need to reward everyone for hiring and mentoring; make it part of performance expectations for all.

What happens when women turn down team housework? When women turn away office housework, they risk being seen as uncooperative for refusing to do work that men are rarely asked to do.[31] But when men volunteer for housework they are recognized and are more likely to get a promotion.[17] Interestingly, total remote working during the pandemic may have helped disrupt these expectations for women. When everyone on the team is remote, everyone is making their own coffee. Taking minutes is done differently in a video call with collaborative tools to capture the team decisions. Gail's manager ran the sprint meeting capturing all the notes and decisions himself. Taking notes in Gail's team has been elevated to a power position, owned by the manager.

Remote work also generates new technical tasks like managing the team collaboration platform, which may be seen as more challenging and important to enabling team success. Managing tools is a necessary component of any product development work. Similarly, bug fixing is always necessary and often seen as boring. One manager was lauded because of her solution to bug fixing; she created a BugFest. Over the course of several days everyone on the team worked the bugs together. The manager provided food and fun and fixed bugs herself. Everyone on the team did the grunt work together, further building team commitment and cohesion.

Whatever the grunt work required to keep the team going we must be on the watch so that women are not doing more than their share. If women are asked to do team housework or

Too often women do more than their fair share of housework tasks to keep the team running

less important tech work too often, they will not get the stimulating work they need. Equally distributing "grunt work", social planning, minutes, and team maintenance sends the message that everyone on the team is of equal value. Everyone must share in doing all the work necessary to maintain the team. Then everyone will get challenging work and maintenance work. Equally distributing challenging and grunt work ensures that women will get the hard assignments so critical to retention.

Given these hurdles to women entering a career in tech, women who have chosen the field must have made a deliberate choice to do a technical job. So, it is not surprising that the Stimulating Work experience is required to hold on to good people both individually and as a team. But because of unconscious bias women may not always or often get asked to do challenging tasks. Managers, team leads, and co-workers may not trust women to do the challenging work. This undermines her sense of value and belonging—and therefore overall team cohesion. Women may also undermine themselves by taking on more team housework than is good for her career. When done well, equal distribution of hard tasks and team housework communicates that we are all "in it together" doing all the work necessary to make something that matters. Managing for retention means managing the distribution of challenging tasks.[21]

Getting challenging tasks is critical to women. But if women think they are less competent, they can doubt their skill and may shy away from challenging task. Our research has also found that women may not be comfortable asking for a challenge. But if they are offered a hard problem and supported to succeed, they rise to the challenge. We explore this experience in Chapter 3, The Push and Support.

CHAPTER 3

The Push and Support

I told my husband that I was interested in a management position, but I didn't do anything about it. We both work at the same company. He is always looking for his next position, so he looked for me as well. He found my current management position and it appeared to be closed. But my husband encouraged me to call the hiring manager and talk to him anyway. He sat with me while I called. When I shared my qualifications, it impressed the hiring manager, and he reopened the position. They hired me in a week. Amber, Research and Development Manager

Women often advance when they are pushed into a challenge and are supported to succeed by family, teachers, managers, and co-workers. Many women hesitate to ask for a challenge or the next promotion. They may not feel skilled or comfortable with self-promotion. They may look at a job description and decide they are not qualified if they do not meet every criterion. Or they may think they need more training. They may not create a plan to advance their career. But women rise to the occasion if pushed into a work challenge or a promotion by someone they trust. And if they then get support from others to plan, strategize, ask questions, and falter, they succeed. Risk is easier when women know they are "not in it alone."

In our opening story, Amber's husband pushes her to apply for the job he knows she wants. By finding the job opening, her husband presents Amber with a challenge—apply for what you want. But her husband also supports his wife every step of the way. He stands by her while she makes the application, strategizes what to do because formal recruiting had closed, and encourages her to call the hiring manager. Amber gets the job, not just because she was qualified, but also because her husband pushed and supported her to achieve her goal.

The Push and Support when received from trusted people, help women take on a stretch goal. It helps women overcome the hesitancy or self-doubt women expressed to us (See Voices on self-doubt). More senior women who have already proven their worth in the industry and to them-

---

In Their Own Voices: Self-Doubt

I have doubts that I'm doing a good job now. I'm a terrible negotiator and I didn't ask for a promotion even though I'm way overdue.

Men don't worry about whether they are ready for the next move in their career—women do.

Women tend to underappreciate themselves. They are less willing to put themselves out there. I see this quality in myself.

I was asked to be a product manager, but I felt that business is a secret language. As a software developer I'd have no credibility. So, I wasn't comfortable taking it on until I got an MBA.

selves are also more likely to take on risk without a push. But women who are just starting out in the field may not yet have a strong sense of their competency or a supportive set of trusted people to guide and coach. So early career women, those in the first five or so years on the job, naturally have more self-doubt and hesitancy to take on risk.

Gender expectations related to self-promotion and being perceived as too aggressive may also play a role in fostering the hesitancy and self-doubt that women must overcome. Also, our survey found that about half of the women "don't know what to do to be successful." Without clarity on what is needed for success, hesitancy about promotions and qualifications is understandable.

For many of us, the need for The Push is not a comfortable finding. Women don't want to think that we need to be encouraged, even shoved, into risks. One senior software developer told Karen that she didn't like the idea that women need to be pushed into challenges but recognized it as her own experience. Whether we like it or not, women shared stories of their experiences of The Push and Support over and over. The Push helps women take on challenging projects and seek promotion; it helps them take on risk. Moreover, The Push and Support are important for retention. In line with other research,[25] our survey finds that experiencing The Push and Support correlates with whether a person is thinking of leaving their job. So, understanding what women need to take on risk is critical to growing women's success both at work and as leaders. Let's explore this.

THE PUSH

> We were meeting with a room full of male software architects. Hugh was also an architect, and we were tasked to co-lead an integration effort. Hugh had credibility and knew them all. I had no technical background and the field of HCI was nascent. After introducing the project, Hugh passed me the whiteboard pen and said, "Karen will now tell you what we are going to do." I didn't know he was going to do that. I had no choice. I stood up and started talking and leading the meeting. Karen Holtzblatt, Co-author

Karen tells her own story about being thrown into the ring by Hugh, a developer and her future business partner. Like Amber's husband, Hugh presented Karen with an opportunity which came with an implicit message of value. The Push is not simple encouragement or gentle nudges. The Push is more demanding, even assumptive: "Of course, you will do this hard thing; I see this ability in you." Women we talked to are encouraged, nudged, and strongly directed by family, professors, co-workers, and managers to take on a challenge.

| *Women's socialization can discourage risk taking* |

The Push and Support factor suggests that women may need to be managed differently than most men. Not because women are genetically different than men, but rather because women are socialized differently. Decision-making and risk-taking take place in a social setting[14] guided by parents, teachers, a peer group, and managers. Societal expectations impact how boys and girls are socialized.

We expect boys to take on more risk, and succeed.[19] Mothers underestimate girls' crawling ability and overestimate boys'.[22] We teach girls to consider the needs of others and to avoid inequality,[1] which makes them less likely to try to get something for themselves. But research shows that women's risk-taking is affected by who they are with; women are more willing to take risks when they are in groups with more women.[3] There is simply a huge amount of "gendered air"[35] that women breathe, both at home and the workplace, which affects their risk-taking.

So, it should come as no surprise that women may not approach asking for challenges and self-promotion the same way that men do. Indeed, company policies that expect women to self-promote for advancement may be using the wrong approach if they want more women managers.[9] But from childhood, through schooling, and into technology jobs, we have found that women will take up a challenge when it is offered and supported by trusted people.

FAMILY AND THE PUSH

My parents were science teachers. My father taught summers in a camp so every year we explored the world—archeology digs, or some other exploratory science project. The family value was to be fearless and just go figure things out that you didn't understand. So, I became comfortable with science from a very early age.
Carol, Director of Engineering

Carol tells us how her family culture taught her to be fearless in the face of science and any challenge. Family can start the pattern of pushing women toward taking on challenges and an interest in STEM.[31, 34, 36] Many women we talked to, like Carol, have parents and siblings in science and tech. (See Voices on family influence) These family members become role models both piquing the women's interest in the field and modeling that an interest in tech is both acceptable and to be encouraged. Indeed, family members are important gatekeepers of a girl's path toward STEM careers.[29]

But beyond modeling and exposing their daughters to STEM, the family can also apply a strong push into tech. Francine tells her story about how her father deliberately influenced her career path.

> **In Their Own Voices:**
> **Family Influence**
>
> *My dad has a Ph.D. in math and my mother is an electrical engineer. An advanced degree was important to the family culture and influenced my choice to be a research scientist.*
>
> *My mom was a psychologist, and my stepdad was a computer scientist. My stepdad encouraged me to be technical. When I interned at his workplace, I got hooked on video games. As a UX researcher, I ended up doing something kind of in-between them.*
>
> *In my senior year of college, my father gave me $1,000 and told me to go buy a computer and play around with it. This was a turning point for me. I liked the computer and I liked discovering ways to use it for school and fun*

My Dad is the only one who can talk me into or out of things because he understands how I think. I played flute and piccolo in high school. I wanted to pursue music. But my father told me I didn't have the patience to be a great musician. I was also really interested in math and science. So, he encouraged my interest in CS in high school and here I am today. Francine, Managing Director of Advanced Technology

Francine's dad deliberately pushed her into a career in math and technology. There is a big difference between choosing a path into music vs. a STEM career. Francine's dad was not a neutral player, letting Francine choose her own path. He deliberately intervened in her choice and she listened to him. Francine and Carol are very successful in their careers. Unlike more classic socialization for women, their families made high expectations and STEM careers appropriate for women.

Families may model STEM as an acceptable career because of their own choices. They may create a culture of fearlessness, challenge, and science. And they may actively encourage and strongly push an interest in STEM. But in all cases, they are also there to talk, share, and generally support that choice. Perhaps experiencing the dynamic of The Push and Support within the family forms the basis for women's comfort when being pushed by other trusted relationships—or taking on the risk by themselves.

PROFESSORS PUSH FOR EXCELLENCE

I was really influenced by two male professors at college. One was the scary Dad who pushed me to improve my skills, like statistics. The other was the nice Dad who was supportive and encouraged me to grow. Eve, Senior UX Researcher

Professors can take up the role of The Push and Support once women go to college. In calling her professors "Dad," Eve tells us that she experienced her demanding and supportive professors like Dads because they played a similar growth role as her own beloved Dad. We view Eve's words as a show of affection, that being a scary Dad means a loving father with high expectations. Similarly, Jane uses the word scary to talk about her experience of being pushed by her professor.

My professor asked me to do an interview with the BBC. I did it because telling him "No" was more scary, than doing the interview. He supported me and believed in me enough to ask me to do it so I could not say no. Jane, Software Developer

Jane rose to the occasion and gave the interview. Because her professor believed in her, she did something scary to her, something that was a stretch, that she didn't think she could do. Scary is another way to talk about how we can feel when someone we trust and value has high expectations for us. These professors' matter-of-fact style when asking women to take on difficult tasks also communicates their belief in their students' value. Within trusted relationships what appears scary becomes possible. The valuing statement implicit in high expectations helps women change their expectations of themselves. These can-do messages counteract more traditional socialization that undercuts women's sense of competence.[8] Below, Joy tells us about her male Ph.D. supervisor who simply expected her to excel—and so she did.

> *My Ph.D. supervisor got me out of my comfort zone. He had high expectations for performance and treated me as an equal in the team. He didn't hold back straight criticism and rolled his eyes if I whined. I felt like he treated me just like a guy—straightforward with high expectations to succeed. Joy, Director of Consumer Insights*

In her perception, Joy's professor treated her exactly the same as all the male students on the team: high expectations, straightforward feedback, and no tolerance for complaining. To her, this was a good thing. The professor's eye rolling in the face of Joy's whining simultaneously dismissed her dislike of criticism and communicated that Joy can indeed take that kind of feedback. His high expectations in the context of their trusted relationship communicates his unshakable belief that Joy could of course do the task.

Whether professors are women or men, they can communicate to women students that going for a stretch goal is expected and that they can do it. The professors stand firmly for women's competence, even in the face of her own self-doubt or the undercutting stereotypic messages from society. They teach risk-taking, competence, and self-confidence all of which are necessary for work success.

MANAGERS PUSH AND ENSURE SUPPORT

> *My new manager encouraged me to facilitate the requirements workshops with our clients. It was a complex software project and the client was difficult so I was pretty insecure. But my manager told me to make a facilitation outline and we talked it over before each workshop. She came to first few workshops; I knew she had my back if things got out of hand. Eventually I told her she did not need to come any more. Sylva, Software Engineer*

Family and teachers may set the stage for women to be responsive to The Push. Then managers in the workplace take up the role. Sylva was hesitant to take on the challenge of standing in the front of a room of difficult clients and helping them come to agreement. But her manager nudged Sylva into learning to facilitate and expected her to take on the work and succeed. Then the manager does not leave Sylva on her own, she helps her learn to do the work until Sylva could work on her own.

The Push from trusted relationships can overcome self-doubt

The Push always communicates that we can do more than we might think we can. It simultaneously communicates value and belief in the woman's skill while asking women to stretch beyond her own self-expectations. In this way The Push builds self-esteem and undermines self-doubt. But there are two sides to The Push and

Support experience. Women managers also tell the story of The Push from the other side—helping their women direct reports stretch. Robin shared how she deliberately built the self-esteem of a woman employee who was having a hard time fitting in with her group.

> *One woman I managed needed a win; she was getting a lot of negative feedback from her project team. I wanted to give her a project that would be all hers and one she could succeed at. I asked her to write a grant. For this, she didn't have to collaborate with the others. She succeeded and got the grant. Then I wrote an email to the team and the other managers sharing her success. Robin, Engineering Manager*

Robin saw that her direct report needed a win and figured out how to give her a challenge which was both a stretch and a likely success. Robin implicitly understood that she could build self-esteem if she offered a reasonable challenge. And because Robin was trusted by her employee, she took up the challenge.

The Push always happens in the context of a valuing relationship with someone who really knows us. This valuing relationship is critical to women if we want them to take on a challenge. When women feel that their trusted person has their best interest at heart and know their real capability, they can move past their hesitancy. Whether delivered by valuing family, professors, managers, or co-workers, women listen and seriously consider the recommended action. They let themselves be pushed into career enhancing choices.

GIVE ME A BIG HAIRY PROBLEM AND LET ME GO

> *A blank sheet of paper is never hard for me. In my first job after school, I worked in a big data warehousing company who just bought a contract with a new leading edge communication product. No one understood it and the other developers were snobs because they said it wasn't real coding. But I just jumped in, learned it, implemented it, figured out what would sell, wrote proposals, went to clients, set the pricing, and made the company a lot of profit. My boss said, "Where did we ever find you?" Robin, Engineering Manager*

Not all women need The Push. Robin loves a blank piece of paper; just throw her into a challenge and give her free reign to make things happen. Robin dove into the problem and figured out everything that needed to be done top to bottom. Robin thrives with and chooses big challenges. Remember Robin is also the woman who thought it would be fun to take a job trying to rejuvenate a dying internet company. But in all of her stories, Robin had managers who believed in her, let her go and provided support when needed. Robin may not need The Push to take on a challenge, but she did need managers who let her solve them and provided support.

Similarly, Carol in her early career, took on every challenge her CEO threw at her but also had support from both the CEO and the men she worked with. She tells her story.

One of my first jobs was working in a telecom company in Africa. They had a lot of military-retired male employees who tended to lead with the respect and restraint expected in a military culture. These men treated me like their own daughter, encouraging me and explaining things when I asked. The CEO wanted to explore all the new security technology. He bought it whether or not the company needed it. He would bring it to me and ask, "How does this work?" He did the same thing when mobile devices were just coming out. I have never been afraid of figuring out technology. In my family no one ever said you couldn't do what you set your mind to. Carol, Director of Engineering

Carol told us that she comes from a family of scientists who taught her to be fearless in the face of a challenge. She brings this attitude to work, and it is rewarded in her first jobs. Again, we see the family metaphor used to talk about her co-workers. To her, being treated as a daughter means that she is encouraged to jump into problems knowing that her questions will be answered and encouraged. When the CEO wants someone to figure out new tech, Carol jumps right in. Bolstered by her family culture and the supportive company culture Carol's natural propensity to be fearless flourished.

Both Carol and Robin naturally push themselves. Both of them also found themselves in a corporate culture that valued their initiative. Both received great support and respect from their managers and the people they worked with. This supportive work context may let their existing inclination to risk blossom. Or perhaps a supportive environment can encourage women to take on more risk. On the flip side, we know that if women's initiative is met with disdain and discouragement as is often the case in tech, their confidence and fearlessness may not persist.

In our survey, 75% of women said "I've always had a lot of self-confidence" which may be an indicator of their willingness to spontaneously take on a challenge. Also, 58% of early career women strongly agree that "I love taking on new challenges." Some women push themselves; they see themselves as naturally confident. But 42% of them do not. Experiencing hesitancy to initiate a challenge may not be true for all women but 42% is quite a large number.

Early career is the most vulnerable time for companies to lose women in tech. Early career is when women are learning what it takes to be successful, the rules of being a professional, and where and how they can jump into a stretch goal. This process of gaining professional maturity on the job lasts much longer than the first couple of years in the industry. So, if 42% of those in early career feel hesitant and self-doubting companies have a retention challenge.

Early career professionals are the most vulnerable to self-doubt

The Push and Support factor tells managers to pay attention to whether women are offered or are spontaneously asking for challenge, to see if they need The Push. Then to provide support whether they were pushed or signed up on their own for a challenge.

But hesitancy, as we said, is not just about needing to gain experience and build self-confidence. Unfortunately, current research still shows that women in tech face unconscious assumptions about their capabilities and expectations surrounding assertive behavior. Perhaps when men in tech interact with enough women like Robin and Carol, men will have their attitudes reshaped by these interactions. But gender attitudes and expectations are hard to dislodge. So these societal beliefs likely contribute to women's hesitancy to take on or ask for hard problems and promotion as we discuss next.

GENDER AND SELF-PROMOTION

By virtue of holding stereotypes—as we all do—women are judged according to different standards than men. Self-promotion is simply more normal and acceptable for men than for women.[27] Gender stereotypes dictate that women should not advocate for themselves, instead they should be communal.[28] Because these stereotypes are ingrained in our culture, there are good reasons that women do not promote themselves. Even Robin who usually welcomes a challenge, avoids self-promotion.

> My strategy when I get bored is to just change companies and find something better. I see my male colleagues self-promote and get a better job inside the company. But I'm not comfortable with self-promotion so I just change companies to negotiate something better. Robin, Engineering Manager

Norms for male and female behavior are unconscious and conformance is expected by both men and women. This is the likely cause of Robin's discomfort. Robin avoids self-promotion by changing companies—a loss of Robin's skill and a blow to retention for the company. Whether she knows it or not, Robin conforms to gender norms. Women know that just like the violation of other norms, the violation of gender norms has consequences.[12] Women who self-promote for a job,[33] who exhibit pride,[4] and who just say "no" to a request to help another person[15] can get backlash.

Women must meet the requirements of the job while navigating hidden expectations for women

(See Voices on backlash) When women advocate for themselves or behave in a dominant manner, they may receive rejection, disdain, and other forms of social disapproval.[26] So, women, knowing these possible reactions, worry about how they will be perceived if they stand for themselves.

Our interviews with women show us that women are not beyond these sex-role expectations of appropriate female behavior. Even worse, women do not have to actually display this behavior. Simply being successful in a traditionally male field like tech undermines a woman's likeability and makes her seem selfish and pushy.[16] For women, behaving "too masculine" has negative effects—but at the same time, displaying traditionally "feminine" behavior can also backfire. If women help others, take on service roles, or facilitate collaboration instead of taking on leadership, they are liked but they are not respected.[13]

Too often, women have to decide between being liked or being respected. This choice is detrimental to their careers; professionals typically need to both be liked and respected.[11] But men do not face this dilemma.[12] Women must walk a tightrope between being seen as too feminine to be productive and too masculine to be likable.[38] Sylva's sorry of her team lead is a good example.

> I had a co-worker who became the team lead. She had a very "feminine" leadership style. Because of that, there were times when team members simply ignored what she said. But then she would come back to it and make it clear this was something that needed to be done. She was not aggressive or unfriendly about it, but the team acted like she had an evil hidden face that was suddenly showing. *Sylva, Software Developer*

This team lead tried to balance an acceptable "feminine" style of leadership with "male" straight talk and clear expectations. She got the work done, but the team did not like her; evil is a pretty extreme characterization of standard good project lead practice. But again, and again, women experience that they are held to different and higher standards of behavior. Behavior that would be judged as assertive in a man is more likely to be judged as overly aggressive when performed by a woman.[32] While men are measured by the requirements of the job, women must meet the requirements of the job but also navigate shadow requirements: differing leadership and interpersonal expectations, devaluing assumptions about her skill, and the need to walk the line on these and other expectations.[23] Expected stereotypic behavior causes women to overfocus on how they interact; it gets in the way of focusing on getting the job done.

If women are hesitant to ask for a challenge, advocate for a job, or take on a direct leadership style, we should not be surprised. Gender expectations create the need for The Push. But because of backlash and because it is simply best practice for growing professionals, women also need support. They need to hear and believe in the message of "You can" and "I'm in it with you."

In Their Own Voices: Backlash

After my training as a Product Owner, I told my manager that I would like to get the next position that opened up. They gave it someone else. That was fine, but later I got the feedback that I was being too pushy about it. What's too pushy about making yourself available for a position and letting people know you are interested in it?

In the past I got feedback that I am arrogant, so I try not to be that way. But my company's promotion process requires me to be more of a self-promoter and I'm trying to adjust to that.

I'm in the highest tier of Help Support. I used to be afraid that if I pushed back, like saying no to tasks people asked me to do that were not my job level, everyone would see me as stuck-up or bitchy. So, I didn't do it. But now I've started to be more confident and I'm pushing back.

SUPPORT

The Push and Support go hand-in-hand for women to thrive in technology. Robin and Carol had tremendous support from their managers and co-workers. Sylva's manager pushed her to facilitate a workshop but co-planned it and was present at workshops for backup. Amber's husband pushed her to apply for the job and helped her with the application and planning the discussion. Support turns perceived barriers into opportunities because someone we trust has our back. This is particularly important for lone women on a team, as Beth shares.

> *I was the only woman on the team at a previous company. My VP of Development was a strong supporter. He was smart and a joy to work with. He always had my back and stood up for me with the outside vendor engineers. Beth, Director of Product Innovation*

Beth experiences herself as "the only woman" plus she did not have a CS degree to give her credibility so she could easily be discounted. But her VP stood up for her to make sure she was heard and could be effective. He was "in it with her." Because of him, Beth was not in the challenge alone. This male manager balanced the impact of being the lone woman.

Both men and women can be supporters—what counts is that women, indeed all employees, get support. In many of the stories we have told so, far more men provided the support. Today, Anita Borg and other women's organizations emphasize the importance of male advocates for women to succeed and be promoted.[7] But research also tells us that people are more likely to receive support from those who are similar to them; people feel more supported when they have others of the same gender on their team and in the company.[20] With so few women in technology or on the team, this may not be possible.

Support means taking the time to build a relationship

Support comes in many forms: someone to bounce ideas off of, answer questions, validate skill in the face of self-doubt, advocate for promotion, champion ideas, and more. Support from partners as in Amber's case is also extremely valuable for women's success.[10] But in the workplace, support is most powerful when it comes from the managers and co-workers who we trust and can ask for help.[6, 17] At the crux of support is that someone takes the time to make a real relationship where women feel known and heard—the basis for trust. These relationships come in different forms as we explore next.

COACHING

> *I took a job with this start-up during the pandemic, so everyone is remote. I mainly work with my manager who is very supportive. When I have a question, I message him and if he doesn't have a busy sign, I ask if we can talk. To make himself more available to everyone he started to have open door meetings a few times a week. But I'm sensitive about asking questions in*

front of others when I'm not an expert. In the open-door meeting, I feel like others are waiting behind me, that I'm taking up time. When we talk 1-1, I start with one question but then realize I have related questions and ideas to discuss. It is more of a conversation and more of an "intimate" interaction. I don't feel rushed. Gail, Software Developer

Coaching is one of the most important kinds of support women want and value—especially if they are new hires. Here, Gail exhibits the self-consciousness about taking up time that we have found in other early career new hires. Gail is trying to figure out the work, the system she is working on, and how to get answers without pelting her manager with small requests. Gail's manager uses an office hour model to support the whole team. But what Gail values is when he takes time to let her explore her topic. One-on-one conversation gives her the information she needs and also builds a relationship of trust. So when later he asks Gail to become a project lead, Gail seriously considers it and decides to take it on. Gail was not thinking of being a project lead; asking her to do the job became The Push. Establishing a relationship of support made saying yes probable.

Randi, a more experienced worker, tells a different kind of coaching story. Randi explains how her manager helped her to become a product manager. Randi has an MBA and marketing experience. When her manager asked her to become a product manager, Randi had no idea what she was supposed to do.

When I took on the job as a product manager my boss was very supportive and coached me. She said, "I'll teach you and help you get started." She told me to work on little things to start in order to gain confidence. Because my manager had the job now, she told me to shadow her for several weeks. She sent me to some product management training, told me to join a book club on product management, had me talk to peers at lunch, and then asked me to run the team meetings saying, "I'll jump in if you need me to." Randi, Product Manager

Randi's manager asks her to take on a big challenge. But in the same breath she committed to supporting her through it. "I'll teach you and help you get started." The manager provides learning resources. She also steps Randi into the job through shadowing, partnering, encouraging questions, and assigning easy tasks to start. Stepping people into the work is an effective strategy whenever people take on a new role or a new job. Stepping people into success is a key technique for onboarding new hires, as our onboarding research revealed.

Coaching is a critical relationship for women in tech, which we will explore more in Chapter 4. Whether from a manager or co-worker, if we want women to take a stretch goal, or the challenge of a new job role, the Push into that risk must come with Support that often looks like coaching.

GUIDING

> *When I'm looking at a technical issue with teams, I take into consideration where everyone is in their learning of the technical space. When a technical issue is raised, I can look at it and know the answer right away. But that won't help someone else know what to do the next time. So, we look at the log together and I walk them through a series of questions to reveal the answer and how to think about the problem. I learned this Socratic technique from my father who was a schoolteacher and taught me that way. Carol, Director of Engineering*

Carol is guiding her teams to learn the technical area. Carol is respected for her technical expertise. But when a technical question is raised, they looked at the log together. Rather than give them the answer she asks a series of questions to reveal the answer thereby building the team's knowledge and confidence. She knows for the team to advance they need to think things through for themselves. Her Socratic approach gives the team space to think and learn. Guiding requires time, patience, and a commitment to the success of the less experience person. Similarly, Helen uses guiding to deliberately build self-confidence as she works with a product owner. Helen is responsible for improving the processes for very large initiatives with problems. She works remotely.

> *On this new project I'm in the listening phase to try to understand the process issues and resource needs. I met with a key product owner to understand what is going on and how to increase their success. This woman has been raising issues for some time, but no one would listen. I think it's because she is both a woman and not a native of the country she is living in so I'm committed to making sure she is heard. I displayed a diagram representing key components of an organization and we brainstormed how to change the current structure. I asked her questions, and we changed the document as we talked. I wanted to give her space to get her ideas out. Helen, Program Manager*

Helen's job is to improve the organization. But she is a long-time successful woman in tech and sensitive to women's issues of being heard. Her tactic to get the best information is also to use guided questions to get to understand what should be done. Along the way she positions herself as an ally to improve the organization and to take the product owner's ideas seriously. Helen is not a people manager, but she is a senior co-worker who takes the time to do her job and support the product owner.

Experienced women already know their own worth. They have been in the industry a long time and have proven themselves time and time again. They are the kind of experienced people that women need to work with to gain guidance in how to be successful. They use a guiding technique to both get the work done and to grow the skill and confidence of less experienced people. This kind of support, guiding the work, is also critical for women to take on challenges and succeed.

CHAMPIONING

Three months into my new management job I had to give negative feedback to a member of the team. Everyone agreed he was brilliant but angry and disruptive. Over multiple conversations, I discovered that he was angry because he felt powerless: "I don't know what to do. I can see all the ways that the organization is doing things wrong and I know how to fix them, but I can't get anyone to listen." I told him I would help him be heard but we had to make a tangible plan, which we did together. I took the plan to my manager to talk it over. My manager didn't tell me no, but he did tell me to expect resistance. Then my manager helped me talk to the next level of management and we argued for the plan together. That senior manager said "Cool, make it happen." We were all surprised. So, we went forward with the plan working together. It was exciting to see the developer do so well and the success we had created together. Linda, Senior UX Manager

Everyone needs people who will champion their ideas and help make them happen. Linda both champions and partners with her employee to make his ideas happen. She validates his goals, helps him make a plan, coaches him in how to present to more senior managers, and advocates for his ideas to even more senior managers. But Linda's success could only happen because both her immediate manager and her senior manager supported her desire to take the company in a new and potentially risky direction. The result was a very successful new direction for the company gleaned by working with a previously problematic employee.

For women to put their ideas forward, they need champions. Linda champions the ideas of her developer. Then her managers champion Linda effort to put forward the proposal. Linda's story shows us that supporting new ideas includes developing champion relationships up and down the organization. If organizations want its people to take risks, if they want to encourage women to take risks, they need to listen, provide support, and tolerate faltering at each level in the management chain.

On a smaller scale, we have found managers showing value by championing women's ideas and showcasing their work to their managers and in group events. Sometimes they advocate for a promotion for a valued employee. One research scientist shared that her manager worked to get her promoted, even though her research wasn't for a shipping product which makes promotion harder. But whether a large organizational change or making sure women's ideas and work are recognized, championing is essential for women to go for stretch goals.

CHALLENGE WITHOUT SUPPORT

In my first research job after my Ph.D. I was told "Go invent the future." But I didn't know what to work on or how to meet their expectations. I did some work and then shared it with my manager. He just kept saying, "Keep going, the work is good." But I felt at sea the whole time with no one to talk to about what I was doing or what was expected. I felt like I was building a house of cards and when it fell it would be my fault. Bonnie, Research Scientist

Taking on a challenge requires adequate support. But what if the manager pushes and then like with Bonnie, no one supports her? What if a person takes on a challenge and is then criticized? If support is lacking, women may flounder. Uncertainty and self-doubt may take over and undermine her growth as a professional.

Bonnie was told to "go invent the future." Bonnie did not know what was expected or what success was supposed to look like. Being told she was doing a good job and to keep going, was not

The Push works best when women can strategize with co-workers

support. She needed someone to talk to, a way to clarify her ideas, and to be able to ask questions. She needed coaching, guidance, and someone to partner with at least until she learned the job. Left alone to face the risk without the necessary support, her self-doubt blossomed, and she became nonproductive. Sadly, Bonnie encountered this situation in each research organization she worked in. She ultimately left the jobs and eventually the field.

When a manager assigns a big challenge to a new hire or someone in early career, they must also plan the support. But managers may not know what to do to ensure early career success. Without the right kind of project and a set of relationships to answer questions, guide, and coach, early career women may not develop the skills they need to be successful. A lack of Support sets the stage for losing women from the company and from the field.

And even with the best intentions, managers may not provide the right support after a Push. Robin tells this story of her worst management moment when she ended up undermining her employee's confidence—the exact opposite of her intention in asking her to talk on a challenge.

A woman product manager who is very introverted wanted a growth goal. So, I encouraged and recommended that she present at an event. After, she came to tell me the story of the conference. She was excited that she had presented and showed me her deck. Instead of telling her she was great I told her that she presented the wrong thing! The PM was crestfallen, and I knew that I totally blew it. I should have told her that I was proud that she did the presentation. Robin, Engineering Manager

Robin's product manager took up the challenge and thought she had a win; she presented, something she had never done before. Instead of celebrating that win, Robin criticized her for the content of the presentation. Clearly, Robin did not provide the kind of support we have been talking about. She did not clarify the content of the talk, review the deck, or watch her present with comments. Robin now knows that her response was destructive. She can't undo what she did but she realized that she could make an amends. Taking on a risk is to be rewarded, not doing it perfectly.

Support implies acceptance that any risk will come with stumbling, failing, and needing to regroup. Tolerance of failure, of a less than perfect outcome, must come along with any Push. And for

> *When we give a Push, don't expect perfection*

success, to build skill and self-esteem, every Push must come with Support. Whether the person who pushes is also providing support or if that comes from someone else, The Push and Support only works if both elements are present. Only then will we build self-confidence and the willingness to risk again.

THE PUSH AND SUPPORT AND REMOTE WORK

> *In my last performance appraisal, I received feedback to speak out more during meetings. I decided to speak up in a meeting where we were asked for ideas to redesign an online survey. I suggested we reframe the questions to reduce the choices for each question to lessen the cognitive load. The VP of Product cut me off and said "That is irrelevant. There's no reason why we need to change that now." The VP has a loud voice, so with cameras off, the comment came out of nowhere. I felt devalued and didn't speak up again in that meeting. As the only UX researcher in the company I thought I'd play a central role in this discussion, so I was surprised by his dismissal. If I were in person, I could have seen his facial expression and how he was reacting like I had in a face-to-face meeting before we were remote. I could have drawn my idea on the white board so he could understand. Kate, User Researcher*

The Push and Support becomes harder when people are remote—especially if they interact with cameras off. Kate has eight years of professional experience but has been at the company for a year. She gets a Push to participate through her performance review. Getting feedback for change in a performance review raises the behavior up as seriously important—it is a strong Push. Kate responds by trying to speak up in a meeting where her expertise is appropriate, but she is shut down. Her own manager did not create a plan or coach Kate in her new challenge. The VP of Product was certainly not informed of her stretch goal nor did he see encouragement as his role. Kate thinks that had she been in person she might have communicated her ideas more easily on a white board. She might have been warned that the VP was not responding well because she could see him. Left with voice only, without support or coaching and without a trusted relationship, Kate was on her own and stopped participating.

This story stands in contradistinction to another new hire's experience with her remote manager before the pandemic. Nancy, who works in the compliance group at a large software company, tells her story.

> *I was asked to resolve a hot issue and felt confident I could do it. But it was in my manager's manager area of expertise and I was anxious about the right approach. I talked to my manager what to do and she told me to talk to her manager and work on it together. My skip manager was great; we worked on the slides together and she came with me to the meeting but sat in the back and let me lead the discussion. I felt very valued and supported.*
> *Nancy, Compliance Professional*

This story is a good example of how The Push may come from one person while support comes from another—even when working remote. Nancy's manager gives her a stretch goal, run a meeting to resolve a difficult issue. Nancy is worried but is guided by her manager and skip manager to success. They provide the support she needs to ensure that the right content and approach are used. Then her skip manager lets her lead but is available for back-up. Nancy takes on the challenge, succeeds, builds confidence, and feels valued.

Our own research and other studies on remote working during the pandemic show that support is even more important when remote. Support seems to act as a "negativity buffer" to help people cope with stress and work effectively.[2] More importantly, support is the most powerful remote work characteristic impacting performance and well-being.[37] In The Remote Work Project, new hires consistently express their need for support from managers and co-workers. Working alone in their homes, they are more likely to feel isolated and may not have attentive managers and teammates available to answer questions on demand.

Remote new hires need support even more

A study at Microsoft tracking meetings inside the company during the pandemic shows that their employees were having more short meetings and fewer long meetings.[30] They were also having more one-on-ones which are needed to develop trusted relationships especially in a remote context. Gail and Nancy got their support through one-on-one discussion, not in office hours with managers, or by raising it in a group meeting. The loss of informal and one-on-one interactions is a continuous theme in our remote work research. Without them, daily support is simply more difficult. Both the women and the product suffer.

Holding remote meetings with video turned off compounds the situation. Too many teams, for a variety of reasons, choose to meet without cameras. Without video Kate lost interpersonal cues and a visual way of communicating her ideas. Kate, like others we interviewed, felt surprised and taken aback by loud criticism devoid of warning or other visual cues from the speaker. As we said, being remote always reduces the interpersonal cues that people use to modulate interaction and communication.[18] And having the camera off signals low engagement;[21] that people are multitasking.[5] With video off, only voice is left. When the communication coming out of the speaker is criticism or anything perceived as negative when taking on a risk, risk-taking will be discouraged.

The Push and Support can work when remote, if managers and teams ensure that trusted relationships are built and if they plan the support needed when taking on a challenge—just like when in person. But when remote it is simply harder.

THE PUSH AND SUPPORT AS A RETENTION STRATEGY

The Push and Support implies the creation of a set of trusted relationships that ensure that women take on and succeed at challenges. Early career women—the most vulnerable from a retention perspective—need support to succeed when taking on a challenge and to become a self-confident professional. Family may start a pattern of encouragement that is then taken up by teachers and professors. But for success in the work-place, managers and co-workers must take the time to build trusted relationships which form the basis for a successful Push because they provide the Support.

The Push and Support may come from different people

In a tight cohesive team, team members are always pushing and supporting each other. The Push and Support is a natural element of a healthy working group. A manager may push, and the team may support. The team may push, and the manager may support. Or co-workers may push and support each other. Indeed, support from co-workers with similar jobs and levels of experience has a great impact on how satisfied people are with their job and how attached they are to the organization.[25] These informal sources of support have been found to be critical for women and more important than formal mentors.[24]

Whether a manager is better at pushing or better at supporting the other half of the equation has to come from somewhere. Once managers know that this is the winning formula, they can make sure that employees get what they need. Given the gender expectations and bias women face they are less likely to feel comfortable self-promoting and asking for challenging work. So, they are less likely to receive challenges. Which means that they are also less likely to get the Stimulating Work they need for retention. To retain women, managers need to pay extra attention to ensure that women take on challenges and stretch goals. But they must also plan the right kind of support to ensure that women succeed. And if women naturally take on or want to take on major challenges, they too need support. They too need support to ensure that biased perceptions of a woman's competence don't get in the way of giving her the big hairy problems.

In some ways implementing The Push and Support is easy. Making people aware of what is needed really works. When Karen gives a talk, she challenges each person in the audience to identify one stretch goal for themselves and someone they should be supporting. Later, women contact her to share their story of how they successfully took on a challenge or supported another women to reach a desired job. Sometimes awareness is enough. But it helps to have a roadmap of what to do to push and support someone to success. The Team Onboarding Checklist in Chapter 7 is that roadmap. It helps managers know what to do to ensure that new hires develop the relationships, get challenge, and learn what is needed for success.

At the root of successful The Push and Support are those trusted relationships. These relationships are formed in everyday work interactions. In Chapter 4, we explore the impact of these Local Role Models.

CHAPTER 4

Local Role Models

I took a job as an assistant professor after leaving my research scientist job at a well-known technology company. I had no experience being an academic. I joined a department with a lot of experienced HCI professors, all doing research, publishing, and going to the major UX conference—exactly what I wanted to do. The HCI lab where I work has five people, two senior and three junior. We were like two dads and three kids. So, I checked in with these senior people to figure out the job. I asked how to manage a large number of students in a class and what percent of students usually fail a test. When I was surprised that three of my undergraduate students didn't get their posters accepted at the CHI conference, they said that it wasn't unusual. He also told me to focus on helping students get accepted in local conferences first and to encourage grad students to pair with undergraduates for more success. Anya, Assistant Professor

Successful women in tech find Local Role Models who challenge, support, and coach them to career success. Like early influencers, more experienced colleagues model good professional behavior and are present in daily life for guidance. These more experienced people also communicate career possibilities through their actions and choices. When the work activities and home lives of senior people or managers look desirable, women may seek advancement. But if managing means long hours, gender wars, power struggles, no family life, and not making products, promotion may be avoided.

In our opening story, Anya describes the kind of relationships women and new hires need to develop into professionals. The full professors Anya works with set the bar for her to reach as a professor: research, publishing, presenting at conferences, and successful teaching. They advise her on best practices for the job based on their experience. These senior professors simultaneously communicate what it means to be an HCI professor and shape Anya into the next generation of professionals. These senior professors are Anya's Local Role Models.

Role models are how we learn. As children, we look at others around us to learn what to be and do. Seven-year-old girls, look to slightly older girls to see the behavior appropriate for them to grow into. For any interest or new life role, we look to those who are ahead

Relationships with more experienced co-workers in the same job are key to women's success

of us in that experience to reveal what to do as a member of that group in society. Whenever we take on a new job, we look at the behaviors and attitudes of those already in that job to learn what to do. And when we consider taking on a new job or life direction, we look to others to see the possibilities and constraints of that role. So too with women in technology.

The role models that influence women in tech the most are their managers and more experienced co-workers. Their actions define the limits and possibilities of that job role. More experienced people doing the same job are even more influential and important. Whether we know it or not, everyone with more experience in the industry, at the company and in that job, are role models for early career woman. They implicitly teach what is expected and what it possible by their behaviors, choices, and lives.

LOCAL ROLE MODELS AND LEADERSHIP

In the 1970s, women sought to change the message to girls about the kind of work women could do. Women were not welcome into most professions and jobs until women started to break down these gender barriers. Women advocated, got, and succeeded at jobs in every profession or career. Every time a woman is the first in her field, she is in the news, even today. One of the wonderful goals of the Grace Hopper conference is to bring thousands of women together to explicitly show that women can be and are in tech. When women are doing every job in enough numbers that it does not make the news, we will know that the door is really open.

Today's women in tech pay attention to the number of women in the company and their job title. (See Voices on women leaders.) When women are not managers, senior managers, board members, C-level managers, senior engineers and the like, a company communicates that these jobs are not really open to women. But when women see other women in the job, they know it is a possibility for them. As we have said, companies communicate whether women are welcome and can advance by the number of women in the company and their job titles.[2, 6, 7] Tech companies know that they need more women in leadership positions if they want to attract women recruits. When women are not in leadership roles potential recruits may not choose to go to those companies.

To communicate the possibilities for women in tech companies and conferences highlight successful women. The Grace Hopper conference includes many keynote speakers by successful influential women to show that advancement is a reasonable goal for women. Companies create events on their campus for the same purpose. Highlighting successful women is important to demonstrate that it is possible to overcome gender

> ### In Their Own Voices: Women Leaders
>
> *When I was looking for a new job, I paid attention to how many women worked at the company. I wanted a critical mass, enough to gauge how as a woman I could advance.*
>
> *At my company they push for women in leadership roles and have a lot of women leaders. One of the CEOs of a partner company is an African American woman. I really like that.*
>
> *At my company, the glass ceiling is at the director level. There aren't many women even in management, certainly not senior management.*

barriers. But it also emphasizes just how unusual it is for a woman to be in that job at all.[9] Heroic singletons can make other women feel deficient as Claudia shares.

> *What drives me crazy is these stories about women who have it all. It does not help me to hear about these amazing women who run an international tech division, have small children, meet with friends, and work out every morning. Since I had children I really do not have a life beyond work and children anymore. So I am doing only a fraction of these supposed role models do and I feel overwhelmed and maxed out. Sure, I admire them, but I can't see myself being like them. Claudia, Tech Consultant*

Claudia is not inspired by the accomplishments of these "amazing women." Instead, she feels deflated. She cannot find herself in these stories. When Ginni Rometty, then the executive chairman of IBM, came onto the stage at Grace Hopper, she communicated that women can indeed get senior jobs. But the distance between her success and an early career woman in technology is long—and there is no map on how to get from here to there. For women to become successful and then advance at work, they need to see women working in their own company, modeling a life experience they can aspire to. When we talk about Local Role Models, we do not mean barrier-breaking women. Rather, women need to see and be supported by more experienced co-workers and managers on a daily basis. These are their Local Role Models.

Research tells us that having a women role model is one of the most effective way to improve women's performance and sense of belonging.[4] Their success shows that women can navigate the real gender barriers and hurdles they face in a technology culture. Men can also be successful role models if they are aware of gender issues and do not reinforce stereotypes.[3] Effective mentoring for women acknowledges and challenges the gender stereotypes in an organization.[5] Organizations can help by raising awareness of these limiting gender expectations. Otherwise, mentors may inadvertently reinforce gender stereotypes. If mentors attribute their own success to adjusting to the dominant male culture, they may push women to conform and perpetuate these gender barriers. Or, if mentors think

Role models aware of stereotypes but who ignore those expectations improve women's sense of belonging

women have "natural" people skills they may steer tech women to administrative jobs rather than promote them into higher technical jobs.[9]

If senior people or mentors dismiss issues of gender, they risk invalidating the experiences faced by most women.[9] Dismissing gender issues, for whatever reason, shuts the door on the discussion of how to navigate a company or the industry as a woman—something all women we spoke with look for. Effective mentors consciously put aside their own limited expectations for women's roles in tech. They respond to the needs of the real individual they are coaching.

The Local Role Model factor highlights that women look to more senior people with the same job type, their managers, and their directors to learn what is possible for their own career.

Every more senior person or manager is a role model, whether they know it or not. Every director or executive manager is a role model for lower-level managers. Women watch how they do their job, and how they live their lives. This always influences what women think the job really entails—and whether they want to advance.

LOCAL ROLE MODELS AND PERCEPTIONS OF ADVANCEMENT

> *I took my job when I saw that the men and women there had kids. They go to work early and go home for dinner even if they work late from home. I felt like I could be a hard, driven worker and fit family into life in a reasonable way. I work a lot and I love tech and a technical challenge. So, if that's what a woman wants, working hard on great challenges and then going home to be with family, then I'm a good role model. Jen, Software Development Manager*

Jen chose her current job because it gave her the life she wanted to live—hard work with a commitment to family, something missing from her previous company. Jen recognizes that how managers balance home and work commitments represents what her life balance there could be in her new job. But even though Jen felt good about her current job, when she looks up to the next level of manager, she does not see something desirable. She shares:

> *It's hard for me to see where I would go after this job. I can't relate to the man in the next position up. His style is not my style and there aren't any women at that level. I have to be able to picture myself in that job and I can't. Jen, Software Development Manager*

Women we talked to need to be able to picture themselves in the next senior role or leadership position. When the more senior job looks appealing, women may seek advancement. But when a manager's work activities and home life looks undesirable, women may not want to advance. Too often women have told us that the life of a manager is unappealing: long hours, lots of travel, no family life, backbiting, and power struggles. (See Voices on managing.) For these women, the culture of management involves activities, decisions, and pressures they do not want to take on. When women look at the lives of the next level up, they see the culture of management that they must participate in to be successful. But they also see the expected home/work balance. When women are committed to a balanced family life, any promotion, any new job opportunity, will be weighed against both the culture of work at that level and the impact on their family life.

Jen took her new job because the people there showed her that a family life was possible in that company. Seeing yourself living your desired life in a job is critical to taking on the next level of leadership. Amber, who has three small children, shares her struggle to see herself as a worker and a mom.

> *I worked all the time for 11 years before I had kids. I knew my company has flex time that allows women to work part time. But I couldn't see myself working with kids until*

I attended a panel at a conference. There women shared how they balanced young children and work. Also, my director who is a woman has kids and works part time. So, I started getting comfortable that I could do it, too. Now a woman who works for me told me that I am inspiration to her showing her that she can have kids and be a manager too.
Amber, Research and Development Manager

Amber believed that hard work and having a family was nearly mutually exclusive. She works at a company with reasonable policies for parents. Yet, she still could not see how she could balance children and the demands of work. Her director reminded her that going part-time for a while was possible, as she did for her children. She gives Amber a Push to be a committed worker *and* a mother. But Amber also needed to hear other women talk about how they work and still have home lives. She needed more of a map to success. Both these types of role models were necessary for Amber to take on being a manager and a mom. Then she, in turn, becomes a role model for others in the company.

If we want women in leadership positions, we need to be sure that the life of a leader is appealing. Women learn what leadership means by watching the behavior of the leaders in their own organizations. If the culture of leadership itself is unappealing, women will not want to be promoted. If women want a family and the real manifestation of home/work balance is unappealing, women are less likely to go after leadership jobs. If they do not have a guide or a map to having a successful life as a leader, they may hesitate. To take it on, women need to know that the life of a leader is livable, maps to their life goals, and how to navigate it.

> ### In Their Own Voices: Perceptions of Managing
>
> *I'm not interested in a VP position. The stress is too high and I'm not willing to put in the necessary effort. I need to spend time with my kids.*
>
> *Executives have to be ruthless. They have to fire people and are typically aggressive. There is too much scapegoating, finger pointing, pressure, and blame. Executives are on their own lonely planet. I don't want to be one.*
>
> *I actively resist managing people. I much prefer building software. People are hard, software is easy.*

The culture of a company is manifest in the daily lives of its leaders and experienced co-workers. What they do communicates possibilities, not what the company says is possible. This includes barriers of bias and gender stereotyping. Unfortunately, the culture of most tech companies works against the company's goal of having more women leaders. To get more women leaders, look at the implicit messages sent by local role models.

LOCAL ROLE MODELS AND COACHING

Local Role Models communicate the possibilities for women in a company. But women also need active Local Role Models to thrive. Local Role Models are the people who provide The Push and

Support and take an active interest in the women's success. Through coaching and partnering with early career women, more experienced managers and co-workers guide emerging professionals to success. Returning to Anya's story about her HCI lab, we see her experienced co-worker professors

> *Local Role Models are committed to growing successful professionals*

guide and coach Anya to success as a professor. She turns to them with questions and for perspective. They take an active interest in her development as a professor.

More experienced co-workers provide the big picture perspective on the work, their role, the organization, and the industry. The developers and designers we interviewed also need to understand the larger context of the work and the organization. We met Cedric, a development manager, in Chapter 1. His new hire, who we also interviewed, shared how she went to Cedric to get a bigger picture of the system and how her piece fit into the overall work of the team. Cedric responded to her questions by providing pieces of information as she pulled for them so as not to overwhelm her. She appreciates his responsiveness; she felt he paid attention to what she needed to become a better professional. Robert took a similar approach coaching his new hire, an early career content designer.

> *The organization is complicated with many layers, influencers, and people who have to buy-in. The challenge isn't the website design, it's the very complicated organization and how to get things done in it. So, I stepped her through it a piece at a time.*
> *Robert, Senior Director of Design*

Robert, like most of the involved managers we talked to, understood that he had to step his new hire into understanding how things work in the complex world of his industry. As a seasoned manager, Robert knew that he was responsible for monitoring and growing his new hire's overall success in the work which included making sure she could function within the existing organizational relationships. Robert also sat his new hire near him so he could drop in multiple times a day to see how she was doing.

Our new hire research reveals that this attention to building connection and providing the information needed for success is essential to the eventual success of a new hire. We discuss best practices and The Team Onboarding Checklist in Chapter 7. Bringing up a new hire first involves coaching competency on the job, successful team integration, and self-confidence. Then after Launch (see criteria for launch), good managers increase the challenges and start facilitating career growth. With any new hire, the onus is on the manager or more experienced profession to answer questions, create connections, and check in enough. These are the Local Role Models necessary to grow new professionals.

Being remote does not change what is needed to onboard new hires. But when managers are remote, attention and planning are critical. Nancy's manager, Helga, was remote before the pandemic. Helga explains how she onboarded Nancy.

I have been a remote manager for a long time so I'm sensitive to how my people are onboarded. When Nancy was hired, I wanted to be sure she was welcomed so I flew us both to her new hire orientation so we could meet in person. I picked her up and we talked in the car on the way to the orientation. While there I made sure that we met the people on her team and the person I buddied her up with for her first project. After she went home, I text or call every day to be sure any questions were answered. We also have one-on-one meetings each week to discuss problems that come up and how well we are communicating. I thought my own onboarding wasn't the best, so I try to be very organized and responsive. Helga, Compliance Manager

Helga is one of our best-in-class managers for new hires. She is very sensitive to what people need for success and connection. She makes sure that Nancy is connected, feels valued, and gets all her questions answered—which Nancy told us she really appreciated. When remote, the requirement for the manager to check-in and be responsive is much more important. Working alone without easy informal ways to get help is an additional barrier to new hires' comfort in reaching out. Nancy doesn't have to worry about that because Helga checks in a lot. Helga also assigned a co-worker who had done the task before to work with Nancy. Then Helga checked in with this co-worker to see how Nancy was doing.

In Chapter 7 exploring the needs of new hires we stress the importance of reaching out to new hires, especially early career new hires, to be sure they get their questions answered whatever resources they need. We also introduce the need for a set of buddies to coach and guide a new hire

> ## What is a Launched New Hire?
>
> **The Team Onboarding Checklist:**
> Chapter 7
>
> - Works well on their own and with others.
> - Asks less "how to" and more "what's this" or big-picture questions.
> - Deliver the work on time with quality.
> - Leverages learnings from previous projects.
> - Exhibits confidence when talking in meetings, one-on-ones, in front of a group, and with an influential person.

to success. Helga helps Nancy succeed by assigning what we call a Work Buddy who partners and coaches this new hire through her first projects. Helga herself is Nancy's Experienced Buddy. The set of buddies we recommend become Local Role Models and the new the hire's first network of support within the company. These buddies are even more important for remote working. Helga provided all of this.

On the other hand, Gail, our remote software developer, was reluctant to ask for help on a project and that led to serious errors. Gail's first Work Buddy and the Experienced Buddy who provided coaching was her manager. But Gail's manager did not check in on her to ensure she was clear on her assignments. Reluctant to waste her managers time, Gail forged ahead thinking she could figure one assignment out by herself. But she made errors and undermined her success.

Our recommendation for new hire onboarding is to follow Helga's model: reach out to them, don't expect them to reach out to you for help. Similarly, Amala, hired during the pandemic, shares her need for a coach and Work Buddy.

> *I was hired to replace another team member and I thought he would be my "mentor" to help me get to know the job. But he doesn't respond to my texts and isn't on IM very often. If he doesn't answer I send an email, but he doesn't answer that either. So for one of my projects, I contacted the VP's assistant to find out who to talk to but that didn't really help. I feel very disconnected. My team members are not used to playing a mentorship role—they just focus on their own work. Ours is a new team and unfortunately my manager is not helping me understand the work or the culture. Amala, UX strategist*

Amala really needs someone to work with. She doesn't understand the job, the work culture, or the country culture where the company is based. She was not assigned a Work Buddy explicitly but expected that the person who had the job before would help. He does not. Amala experiences everyone at the company as heads down doing their own thing and unresponsive to requests. Clearly, no one is checking in on this new hire and ensuring she is supported. Amala is sitting alone in her home. She can't drop-in on anyone, and no one sees her, so she feels invisible. She can't even learn by watching others—she interacts with no one else except in formal meetings, often with cameras off.

Local Role Models are not just the people we watch to see how the job is done. Local Role Models are also the people that grow early career professionals through partnering, coaching, guidance, and championing. To ensure the success of new hires and early career professionals, we need to deliberately design and assign their Local Role Models. Identifying the right set of buddies is always essential no matter the work context is critical, especially when anyone is working remotely at home.

Good Local Role Models are also the source of building trusted relationships. They become potential mentors within the organization and after they leave the job.

FINDING A MENTOR AND BUILDING A NETWORK

When done well, working relationships with what we call Experienced Buddies become trusted relationships, which form the real basis for mentorship. Mentorship involves the implicit or explicit taking on of a career-coaching role for a valued employee.[8] Often mentorship is a reciprocal relationship. The manager or professor needs people to do the work who are reliable, dependable, and can operate independently. When they see that an employee or student meets those criteria—and they click interpersonally—the senior person can become interested in helping with promotion and career decisions. Robin's story reveals this reciprocity.

At my new job I had a woman VP who was interested in tapping into my skill to help reinvent the company. The VP was a wide-ranging thinker and taught me about the Internet of Things, which was cutting edge at the time. I felt like my brain grew in a short period of time. I admired her and she became my mentor. When I negotiated a job change, I talked to her because I had trouble tooting my own horn. She helped me get over the awkwardness of self-promotion. Robin, Engineering Manager

Robin was hired by and reported to this VP. The VP valued Robin, today a seasoned professional, for the skill she would bring to the table. The mentorship relationships blossomed within their daily work. The VP shared knowledge that Robin needed to do the work and Robin fulfilled her promise as an innovative business thinker. The new ideas they generated came from this collaboration. When Robin left that company because it collapsed, the relationship continued. Robin has a mentor that she can call upon for career advice. Similarly, Maya, a database coder turned designer, also first builds a relationship with her manager—then she reaches out to him for career help.

While I was an undergraduate, I worked as an intern at a large retail company building databases and business intelligence tools. My manager was my first mentor and the first person I could talk to about a career since no one in my family went to college. He was really interested in me as a person including how I was doing at school. When I started, he told me, "Right now don't think about what you don't know, just memorize everything and get it done." I followed his advice and learned. At one point I looked at the code and saw that it wasn't good based on what I was learning about interaction design at school. My manager moved the other coder to a different project and gave me the responsibility for the design. He told me, "You are the best UI designer I have ever worked with." So, after the internship when I was looking for a job, I reached out to him for a recommendation and advice. Maya, Interaction Designer

Maya developed a real working relationship with her manager, who provided perspective to help her succeed. Her manager, like in Robert's story above, helped her understand the expectations and culture of this workplace, which every new hire needs. Maya especially needed someone to talk to about career because she is the first person in her family who has gone to college. New hires with family members in STEM professions often turn to them for perspective. Instead, Maya looks to her manager to provide that professional perspective.

But Maya's work also advanced her boss' desire to produce a superior product. He supported her design idea because he believed in her design and her skill as he tells her. So, when Maya reaches out to him for career support, he responds. He fights to open a full-time design position for her within the company, which she took. Maya had found a mentor, and her mentor had found an excellent professional. A mentoring relationship works when it is reciprocal. Experienced Buddies, VPs, managers, or senior co-workers become mentors through a reciprocal relationship where the worker provides impressive value, and the experienced person provides supportive coaching. Through working together, a trusted reciprocal relationship becomes real mentorship.

Women report that they need mentors. In our survey, 81% of women say they have outside mentors and 77% agree that "My best mentors are people in my company I clicked with who know more than I do about my work." But we also find that men are more likely to click with and find Experienced Buddies who become mentors. In the age bracket 25 to 35 only a quarter of

the women but one third of the men strongly agreed that they had clicked with someone as a mentor inside the company. Of women, 49% agree that "I do not have enough people in my company I can really talk to about my career aspirations." Women need mentors or

> *Women do not have enough experienced people to talk to about career*

at least coaching from Experienced Buddies within their own organization.

Mentorship may develop through ongoing interaction as we have seen in many of the stories we have shared here. But companies may also assign someone who is supposed to be a mentor as part of the new hire process. Allison shares her experience.

> *I feel that my first manager mentored me because she was being evaluated on mentoring. I think they eventually stopped evaluating her on mentoring because she stopped mentoring me. I felt cut off. Allison, Senior User Researcher*

Assigning a mentor doesn't mean that the designated person will take on either coaching or mentoring. Allison experiences herself as "dropped" as a mentee, leaving her on her own to figure out the work culture and how to be successful. Measuring managers on mentoring may or may not encourage a real mentoring relationship. Often when companies assign a "mentor" they are really talking about assigning a Work Buddy. When women say they need a mentor, they may, like Amala, be looking for a Work Buddy to show them the ropes. Work Buddies are essential relationships, but they are not ongoing mentorship. At the core of mentoring is a reciprocal relationship between people who click where the experienced person decides to be responsive and to provide guidance in an as-needed way. Clicking can't be mandated, nor can reciprocity, but when it happens mentorship begins to develop. This is most likely when Work and Experienced Buddies have the same job type—whether or not they are assigned.

Women need and look for more experienced in their company with their job type to talk over work and career. Whether assigned or not, new hires desire and need help from someone who understands their work. Julie shares her experience as a recent graduate and new hire.

> *I was assigned a mentor at my first job. He was a man with the senior version of my job so he could help me. He guided me through every step of my first project. Then when I was doing a project of my own, he met with me to give guidance. I still go to him with questions. Julie, UX professional*

Julie's "mentor" was a Work Buddy who was senior enough to guide this new hire's first projects. He was a designated go-to person for questions and perspective on how the UX job was done in the company. Whether he will become an ongoing mentor for Julie is unknown, but he was certainly the right Work Buddy to get her started. If the relationship continues, he may also become an important part of her senior network at the company. But when a manager does not have the same job role, they often do not understand what is needed to grow the new hire professionally. And they cannot help them navigate the organization appropriately for their role.

Our survey shows that participants who had the same job type as their manager were more likely to see themselves "living the career life of their role models; mentors, or manager above me." They were also more likely to have mentors in their company that they clicked with. But what if that is not possible? Robert hired an early

> *Mentorship starts with a trusted reciprocal relationship between people who click*

career content designer, but Robert did not have a background in content design. His new hire was working alone in his group. To help her feel connected and grow in that profession Robert reached out to another content designer with more experience in a different group and geographic location. He asked the experienced content designer to meet regularly with his new hire and help her learn about the job role. She agreed to take on this advisory role. Robert tells us.

> *The two women clicked and set up regular coaching conversations to share their work experience. She was also available for quick questions through instant messaging. And the senior content designer introduced my new hire to the messaging channel for content designers as well as professional events. I was her manager, but she needed someone to talk to about her field. Robert, Senior Director of Design*

Robert knew that he could not guide his new hire in her profession so he hooked her up with someone who could. More importantly, this senior content designer connected Robert's new hire to the wider content designer community. The natural mentor for any employee is their manager. But if the manager and employee do not do the same job, the new hire needs someone more senior who does. To be successful, women need a network of senior people to talk to. They need the right person to talk to about any given situation. One senior advisor, the manager, is rarely enough to build the kind of network of relationships women need to be successful. Critical to that network are people who do the same job.

Everyone, no matter how experienced, needs someone trustworthy with the right knowledge to talk things over with sometimes. Life and work situations arise which require perspective. (See Voices on finding the right person.) To get that perspective women must build a wider network of relationships. Managers can help new hires start to do this. Whether women find these relationships on the job, in previous working relationships, professional organizations, women's groups, or conferences building a network of senior relationships is critical for success. These people may eventually become long term mentors.

Through their Local Role Models, women can be helped to create a larger network of the right experienced people to support career development and to strive for leadership positions. These senior relationships do not always or ever need to be other women. Women in tech are often few, with less power than men. Asking more senior women to take on every woman who comes to the company asks them to spend significant time away from developing her own career potentially with someone who is not a co-worker. And since men often have more influence, early career women need male coaches, influencers, and mentors to champion them.

The importance of mentoring and building network of influencers for women is well documented. Unfortunately, women are less likely to have a mentor. Men usually have several mentors throughout their career, but women often report not having any. Contrary to what was previously believed, access to potential mentors is not the most significant barrier—rather mentors do not gravitate to women because they do not see them as technically capable.[8] Without seeing women's real potential because of gender stereotypes, mentors are less committed to mentoring women. Our survey also shows that in the age bracket 25 to 35, men feel more supported than women: 26% of women but 41% of men "strongly agree" that they are supported by a person in their company who is a mentor.

> ### In Their Own Voices: Finding the Right Person to Talk to
>
> *I have a 9-month-old baby. There are no women in my company I can talk to about how to juggle home and work.*
>
> *I don't really have mentors. I look for people I can bounce ideas off of and get feedback for the situation I'm dealing with. I don't use just one or two people.*
>
> *I needed to understand mergers and acquisitions. My background is in development, so I didn't know anything about it. So, I reached out to a woman manager I knew, and she explained it all to me.*
>
> *I wish I had a mentor who I could trust and talk to about everyday issues that come up. It would really reduce my stress.*

Without deliberate planning, women may not build a go-to network of senior people. They may not get the coaching and perspective they need from experienced people or develop the right set of buddies. They may not have effective Local Role Models. To grow more women leaders, women in tech need Local Role Models living an appealing life, who will take on the task of coaching and mentoring appropriate for that individual.

ADVANCEMENT AND WOMEN IN TECH

The Local Role Model factor gives us insight into why women may or may not want to advance into a leadership role. Women are influenced by the life and work of more senior people, whether women are moving from individual contributor to management, manager to director or director to VP. If the work itself, the home/work balance, and the culture of those in the next level up does not look appealing, women may not want to advance. To succeed and to become leaders, women need a network of local managers and senior people to partner, coach, and advise them. In other words, Local Role Models are essential to developing women into professionals who will advance to leadership roles.

Of course, just like men, not all women want to become managers or take on leadership roles. Early career people are just figuring out their own career and are not necessarily ready to guide someone else. The pull of doing the work is also great; women love making things. If management means only managing people, those who love to make things may not want to give that up. For these people leadership that still includes making products is another way to encourage women to advance.

For many, management is an appealing goal in the abstract. Our survey tells us that three quarters of women ages 25 to 35 want to be managers. While this number is even higher for the men in that age bracket, almost 90%, a large percentage of the women are interested. More importantly, 78% of women can see themselves "living the career life of my role models, mentors, or managers above me." So, we do have a large portion of women in tech who will become managers and leaders if we help them get there. But as we have said, 45% agree that they do not know what to do to become a manager. And 49% say they do not have enough people in their company they can really talk to about career aspirations. Add to this built-in bias in society and the tech culture, and co-workers who doubt their competence. We aren't making advancement easy.

> *Too many women don't know what to do to become a manager or leader*

Women need to know where they stand, the path to leadership, and to be helped to think through a career path. Leadership is hard if women do not know what to do to be a manager or to be promoted. Given stereotypes, seeking promotion will be harder if women are expected to self-promote to go after leadership jobs. And if decisions about who to promote or give challenging work are influenced by biased preconceptions about what women can do, women will not get leadership opportunity.

Corporate policy can also get in the way. Some companies of people we spoke to had created a policy that prohibits managers from suggesting directions for employee growth and advancement. In the name of supporting employees' goals and a self-driven career, they inadvertently removed the guiding hand of Experienced Buddies. People just starting out in the field and many early career people simply do not yet understand the options that are possible. Nor do they have the insight or confidence necessary to direct their own career. The need for women to have a set of effective Local Role Models is clear. Managers and other senior people must take up the job of deliberately "raising" new professionals.

Last, our favorite insight came from Cedric, a manager Karen worked with who used the Team Onboarding Checklist for his new hires.

> *I was good at my job, so they made me a manager. I've never been trained in how to understand or support people—we were hired to get a job done. Thinking about the best way to support and grow people isn't in my wheelhouse. This year I had to bring on three new hires. I don't have the time to do it right. In addition to managing, I'm also expected to code and oversee a larger technology effort. But I've learned from using the Onboarding Checklist how much I have not been and could be doing. So, to make it work I told the team we are all responsible for bringing a new hire up to speed and getting them connected. And this is working, by making it a team effort the new hires are doing much better! Cedric, Development Manager.*

Cedric is willing to learn what to do to better onboarding and develop success in his new hires. But he did not have the time or the skill to do it all. His insight is to use the team. Together they can provide the kind of attention, coaching, information sharing, and partnering that he now knows is needed. The team embraced the approach and collectively supported new hires to develop skill success and connection with the group—exactly what is needed for retention. Every member of a team, from the perspective of a new hire, is a more experienced Local Role Model. Collectively they can "raise" a new professional.

Local Role Models, managers, and Experienced Buddies are essential for retention. They help make women in tech successful when they see their job as coaching women to develop skill, connect to the team, gain perspective, navigate the organization, understand promotion, and build a network of senior relationships. But also for women to advance, Local Role Models must model an appealing lifestyle as a leader. This also depends on the company culture. Many women looking at leadership roles look for the flexibility and space to create reasonable home/work balance. We explore this issue next in Chapter 5, Nonjudgmental Flexibility for Family Commitments.

CHAPTER 5

Nonjudgmental Flexibility for Family Commitments

My son is now 3.5 years old. After maternity leave, it was emotionally really hard to leave him, but I was worried about becoming irrelevant at work. I felt like they treated me differently once I become a mom, that I got less responsibility. I was pumping three times a day and had to leave early. I didn't want anyone to see me leave early so I snuck out and didn't say goodbye. I worried that they now thought of me as a mom and not a respected programmer. So, I didn't talk about my home life or my son. Now I feel less guilty about leaving work for family because my manager is so supportive of family life. And I saw that my co-workers were involved dads so I thought "OK I can do this too." Ellen, Quality Engineering Lead

Women with children thrive when their team and managers make it easy to balance home and work. Women may worry that they will be judged poorly if they have to take time to care for children. And they fear that they will be excluded from challenging work once they become moms. But when both the team and managers work out strategies to make hard work and family life co-exist, they stop worrying and focus on the work.

The Nonjudgmental Flexibility factor reflects the ever-present theme of balancing work and childcare for women with children who we interviewed. Ellen's story is a good example of the myriad of feelings that women bring to the workplace with respect to taking care of children. Ellen is keenly aware of what others think about her as a worker. She did not want to seem unreliable, that she wasn't pulling her weight. Taking time to pump milk and making the pick-up time for daycare cut into the day,

> *Women worry they will be seen as unreliable workers for spending time on family demands—so they hide it*

so she snuck out. When her manager was supportive, when the dads on her team modeled parental involvement, Ellen relaxed. But anxiety about balancing home and work runs deep in women, as we will discuss below, often calling for The Push to overcome it. Ellen also shares:

> *I was worried about taking my maternity leave. My director, also a woman, gave me a pep talk and told me to focus on my family. She and my co-workers would help her through the transition—and they did. Now, if my kids are sick one of my male team members just says, "Stay home and don't worry."*

Ellen's manager and the team make it explicit that dealing with family is a priority, that women (and men) can go home and take care of their kids without guilt or career concern. They give her a Push to go home and they Support her by making sure she does. They flex the work around her situation. Knowing this, Ellen stops worrying and can focus on the work. Because women have baggage around parenting, worrying about what people think and the possible consequences of taking time for family, The Push and Support may be needed to overcome women's concerns. Ellen's co-workers deliberately push aside her fears.

Implicit in Ellen's story is her team's value of and commitment to fulfilling family obligations. For this team taking the time needed for family is legitimate for all workers, balancing home and work demands is a team norm. But also when flexibility really happens in the face of competing demands between home and work, team members with home obligations know they are an important part of their team. Joy shares her experience of this Nonjudgmental Flexibility from a previous job.

> *I was working in a very large multinational company with a high-pressure, high-stakes job. I loved the work and my team. The company was headquartered in Europe where the attitudes, culture, and law are more supportive to moms than in the U.S. When I had my daughter, my male manager and teammates declared that I had "special privileges." I worked at home in the morning until lunch, breastfeeding my daughter two times before I took her to daycare. I worked in the afternoon but took an hour off to pump milk. Then I picked her up at night in time for dinner, nursed again, and worked after her bedtime. I did this for about a year. I felt privileged. I could not have kept working in this kind of job without this support from my manager and my team. Joy, Director of Consumer Insights*

Joy had "special privileges" when she became a mom. Joy's team declared their home/work value: a commitment to making sure that a team member can be both a mom and a contributing worker. They team flexed around her schedule, her daycare, and her breastfeeding needs. Both the

Flexing around women's home demands sends a powerful message of valuing

manager and the team changed their expectations of her, moved around their schedule for when they needed to collaborate, enabled her to work at home, and generally redesigned the teamwork around her needs. The result was that Joy could care for her young child without losing the stimulating work and team involvement she loved.

Joy loved this team; she already experienced it as dynamic and valuing. Flexing to make family life work is related to the cohesion of this team. Nonjudgmental Flexibility is both the result of, and also builds on, a team's cohesion. The reward is that the team gets Joy's skill and gratitude. The work gets done with quality—and so does the parenting. When teams accept and flex around home commitments, they send a profound message of a woman's value to the team. And doing so is also an expression of being a Dynamic, Valuing Team.

Nonjudgmental Flexibility is also related to Local Role Models. If the members of the team and the manager take care of home life, they are models that home and work can co-exist. We have already seen this behavior. Ellen was affected by co-workers who were involved dads. Jen told us that she took the job at the new company because she saw men and women working hard but going home to be with family for dinner. Amber also needed role models to believe she could work and parent. She was influenced by the conference panel on balancing young children and work and her director who worked part-time after a new baby. When the people in a company take time for kids, women learn that home/work balance is a real company value. They let go of their worry and guilt. Moreover, when the corporate policy is very explicit, it helps managers provide the perspective and support women need. Allison, who we interviewed a second time, shares her story. She is now a manager.

> *Our company has a family-first policy and people have been great about making it easy for me to have my daughter. But when her daycare closed for vacation for 2 weeks, I was really stressed. I was pushing forward a multi-year strategy that I'm invested in. Then we had a reorg and I didn't know what would happen to my work. I shared my angst with my manager. He just said "Reorgs always come. Some things we've been working on will get done, others won't. We can move things around now to deal with your vacation. Forget about work, go be with the baby. You won't get this time back." Allison, UX Manager*

Go, Allison's manager says, go be a parent, we will take care of the business. Allison's manager had been with the company a long time and chose not to leave explicitly because of its family-first policy. Not only does he give Allison a Push to go do her life, but he also models this same family commitment. Understanding the organization, he provides perspective on reorgs and helps her let go of that worry; he helps her make family a priority. Allison's manager is both a Local Role Model of committed parenting and gives her a push into parenting.

Companies that explicitly define a good home/work balance policy—even a family-first policy—communicate that women (and men) can both take care of family and have serious, stimulating careers. Explicit and implicit policies at work communicate what is OK to do with regard to parenting. These values play out in the behavior of managers and teams. Eve tells us about the power of being able to bring kids to work.

> *My company is pretty family-friendly. If I have to bring my daughter to work because she is out of school for some reason I can. On some days there are several children at work. They are pretty cool about it—when you are in a bind you can bring your kids in.*
> *Eve, Principal UX Researcher*

Eve tells us that "family-friendly" means workers don't have to hide that they have families; that companies understand that sometimes people need flexibility. Jane tells a similar story of when she worked at a consultancy that met in the founder's home. If they needed to, people brought their kids to work. The work and the kids happened together in the same living room space. When people bring kids to work, they acknowledge that co-workers have outside lives that have to be accommodated. When managers and teams encourage parents to deal with family responsibility or

In Their Own Voices:
Home/Work Balance in the Pandemic

My company understands that employees need to focus more on work–life balance to cope with life changes due to the pandemic. We organize our workday to end earlier. We are understanding when coworkers don't answer email as quickly as before.

My team is good at making an effort not to schedule meetings past 4pm to support good work—life balance now. And we have no lunchtime meetings either.

Everyone is more understanding and flexible about what people have to do to meet the demands of their life. I had to deal with my in-laws when they were exposed to COVID and dropped all my meetings to make this happen. No one complained. But it is really stressful for me, my husband, the nanny and my kids—I feel trapped.

I'm making an effort to stop work at 5pm. I used to work really long hours but now that doesn't really work. I told my manager if there is really an emergency, he can call me on my phone, but I would otherwise be off. He was fine with that. Everyone gets it that we need to take care of things at home. They are tolerant if we need to step away during a meeting but that would be rude in an in-person meeting.

just to enjoy family life, they acknowledge that family is a priority to be balanced reasonably with work. No matter what the company policy toward work and family is, the reality of that policy is played out on the ground in the culture created by managers and teams.

The pandemic pushed companies into a more flexible, tolerant home/work policy. And the work got done. But much of the burden to support children without in-person school and daycare fell to women. Women in tech were nearly twice as likely as men to have lost their jobs or to have had to take an unpaid leave of absence due to the pandemic than men.[28] 5

Yet the people we interviewed consistently extolled the flexibility of team members, managers, and companies to ensure that work and family could co-exist. (See Voices on home/work balance.) Babies attended meetings, cameras were off to enable folks to deal with messy houses and children, moms and dads walked children and listened to meetings, teams ended meetings a 4pm, parents ended the day at 5pm and sometimes were back to work in the evening. And when fathers took up a greater share of childcare, mothers were less likely to suffer a negative career impact.[26]

Certainly, nothing was easy about the juggling required during the pandemic for women or families. But it showed that flexing to family needs was possible in the tech industry, even in the most demanding of companies. Whether this flexibility continues is part of the work experiment we are

---

5   During the pandemic women have taken back gendered activities like childcare, housework, and now overseeing schooling disproportionately to men when we look at the total population.[8] They have also reduced work hours across all job types.

witnessing. If companies can learn that good work can still be done when teams flex to home commitments, perhaps we can ease the considerable stress parents face as they balance home and work.

Over and over our data shows that cohesive, dynamic teams flex to each other's needs and respect each other's outside time commitments. All it takes is a commitment to flex and a strong message to women and men that it is OK to go be a parent. We need to send a strong message to counteract gender expectations, fear, and the very real tightrope women walk to when dealing with demands from home.

THE BALANCING ACT OF HOME AND WORK DEMANDS

When I switched jobs, I was very self-conscious in the first week because I was late scrambling to get my daughter to daycare. I felt like everyone in the room was judging me. So, when I was late to work because of a daycare snafu I made up the time. I could explain what happened, but I still feel like it made a bad impression. Now I have developed a whole network of babysitters, before- and after-school programs, vacation camps, and helpful websites to be sure my daughter is always covered. Joy, Director of Consumer Insights

When Joy changed jobs, like Ellen, she was sensitive to how her new team might perceive her performance. As a single parent, she has all the responsibilities of managing childcare support. Joy did not want to make a bad impression, which is reasonable since in tech, long working hours and being constantly available is often expected.[22] Both Ellen and Joy measure their reliability by how much time they are putting in. Worry for how much time is put in at work comes from society's assumption that women with children do not work as hard as men and are more likely to quit. Sadly, longitudinal research on women in tech in the U.S. confirm this increased quit rate for parents if they do not get sufficient support for balancing work and other competing responsibilities.[1] Data following full-time tech professionals after the birth or adoption of their first child showed substantial attrition. Almost half of new mothers and almost a quarter of new fathers leave full-time employment after having children.[4] Clearly, women in tech struggle with the demands of parenting within the ever-available work culture of many tech companies.

Women are more likely to quit when they have children. But men with children quit as well. Implicitly they both are influenced by societal and gender norms in play. Indeed, the ideal worker norm for both women and men stipulates that a committed and valuable worker keeps their family responsibility invisible at work.[23] But women are also confronted with the expectation of family devotion. This norm assigns women—but not men—the primary responsibility in family caretaking.[3] Motherhood is expected to override all other commitments, implicitly demoting work to a secondary commitment.[29] Being perceived by society or the team as not fully devoted to motherhood brings the sanction of being considered

Moms in tech balance expectations to be devoted parents and ever-available ideal workers

a bad mother—Germans use the term "raven mother" to refer to women who pursue a career while they have small children.[7]

So, it is not surprising that women are sensitive to what others might think about them as workers and mothers. The ideal worker norm and the family-devotion norm are at odds with each other. Complying with the family-devotion imperative, mothers take on more responsibility for childcare. But no matter the stress that implies, they may not talk about their needs as parents or use parental flexible policies that companies offer because it impacts their perception as a "good worker." It is no wonder women feel torn by the different demands of being an ideal caretaker and an ideal worker. It is no wonder they worry.

Research shows this worry is not unfounded, since men and women are perceived differently in their role as parents.[33] For example, when women are just as productive as men, they are seen as less productive if they use flexible family work options.[5] Joy worries as she starts her new job because her new team can see her struggle to get daycare in place when she moves to a new city. She is late, and so by social norms she is revealing that she is neither an ideal caretaker nor worker. She makes up the time, but her struggle is "seen." She fears that her stature as a worker has been impacted. Similarly, Ellen fears that once she has children, she will lose her professional stature— she'll be seen as a mom not a competent programmer. She is convinced that her team will treat her differently and give her fewer challenges and responsibilities. She manages that fear, and complies with the standards of an ideal worker, by not talking about her home life. Eve also fears that any slowing down for parenting will negatively impact her career momentum.

> I have two kids; my daughter is 6 and my son is 18 months. Juggling work and the kids is really demanding now, and I don't know if it is sustainable. But my mom worked when I was growing up and I never questioned that I was going to have a career—if she could do it so can I. So, I'm not willing to give up my career or scale it back. I know I will lose influence and the ability to advance at work if I do. One of my friends has an MBA and stays at home now. She told me that she thinks her kids won't respect her and that when she goes back to work when they are older, she won't be at the same level as those that didn't take time off. My son is little now, so I know it will get better in the future. Eve, Principal UX Researcher

Eve does not want to reduce her career possibilities by slowing down for kids. Eve is following her mother as a role model of handling work and children. Like Joy, Ellen, Jen, Beth, and many of the women we talked to, Eve wants a challenging career *and* to be able to parent. But it is not easy as she further shares.

> Sometimes I feel guilty about not being a stay-at-home mom especially when my daughter says, "I wish you could just stay home!" But we organize our lives to maximize time with the kids. I really value our 45-minute commute because I get to talk to the kids and find out about their

day while I'm sitting in traffic. And we get home early enough so I don't have to rush dinner and bedtime rituals.

Eve wants a career, and she wants to be a good parent. She judges herself on whether or not she is being a good parent; whether she is complying with the ideal caretaker norm. Whether we like it or not, women are still seen as the primary caretakers in most countries.[2] Women grow up with these gender expectations and do not easily shake off the feeling of family responsibility—or the guilt for not slowing down or taking time off. Amber also manifests this internal struggle.

> *People come up to me and tell me that they don't know how I "do it," being a manager and having two small children. I don't really see this as a compliment; I feel like they are judging me, that they think I'm neglecting my kids. I especially feel this when the people telling me are women who took off time to take care of their own children.*
> *Amber, Research and Development Manager*

Even a compliment on how well Amber juggles work and home is perceived by her as a possible judgment of her parenting. Amber is wrestling with her own internal voice judging herself for not taking off more time to care for her children. If she makes a choice to balance career and parenting, even when she does it well, she worries that she is not the ideal caretaker. Women want a challenging career, but they are susceptible to guilt and to feeling judged as workers and as parents.

Women's fear is justified. Research confirms that women are afraid that they will be replaced if they make use of work flexibility.[25] Implicit work values reward workers who support the team to meet a deadline over using flexibility to accommodate parenting even when it is a company policy. The tech culture upholds this tacit agreement that good employees do not jeopardize the coordination of the team effort to deliver.[15]

So, women are caught in a bind. They must deal with their two conflicting self-expectations: being a good parent and being a good worker who delivers on-time no matter what. Even when company policy offers needed flexibility, implicit company or team values work against using that flexibility for fear of negative career impact. In our survey the set of questions for Nonjudgmental Flexibility as a whole do not significantly correlate with whether women are thinking about leaving their job. But two of the survey questions do: "I feel like women are looked down upon by my team if they take time off or work at home to deal with children" and "What-

Using company policy for family leave can undermine career success

ever the policy, my team and/or manager do not support reasonable home/work balance." These survey questions reflect the conflict women with families face. Women feel judged and watched; they rightly feel that their career will suffer if they choose to take care of family needs.

This ever-present inner conflict creates the need for explicit Nonjudgmental Flexibility. Only when the team supports a home/work value for all, the manager gives women a Push to parent, and Local Role Models live lives where home and work are equally important, will women put aside their fear, guilt, and inner conflict. Nonjudgmental Flexibility plays out across all the factors in The @Work Experience Framework. Taken together, addressing women's inner conflict between home and work demands is critical to retain women in tech.

CAREER CHOICES AND PARENTING

Women in tech want a career and women in tech want families. Both children and work take time. All the women with children we interviewed want time with their children—and they want interesting work. Someone has to take care of and be available for the kids. School and daycare do not necessarily cover the whole day or deal with sick children. Work isn't confined to the hours of daycare and school. And good company or country policies only address the first few months or even a year of a child's life. They don't extend that support of parents into the rest of a child's life. So, families need to make choices.

All the women we interviewed find a way to fit family into the constraints of the workday. None of them want to be stay-at-home parents. So, some women choose to continuously balance home and work demands on a daily basis. But others make different career choices so they have more time for family. Often they limit their work hours. Jane shares her priorities below.

> *I don't want work to be too exciting in case it sucks me in. I have to pick my daughter up at daycare and if they call, I have to leave. So, I work a 32-hour week. I don't need more ambi-tion. I want other things in my life than a career. My current job is comfortable, and the work is fun. I'm married. I have people to play games with including my daughter who is a mini-me. I don't want the stress. I think men have it easier. Jane, Technical Consulting Engineer*

Jane is happy with the overall balance of her life for now. Jane loves solving puzzles. But cod-ing can pull her in so that she loses all sense of time and might miss a daycare pickup. So for now Jane doesn't need the most challenging problems. Jane has also already experienced the stress of a high-impact job and a management role before she had kids. She doesn't want that either because she wants to spend time with her daughter. Her work is compelling enough and being a senior individual contributor with a reduced work week fits well with a reasonable family life. Men have it easier, she says, pointing to the different expectations on women and men.

Like Jane, Deb also curtails her work hours to be the kind of parent she wants to be—and so does her husband.

> *Ever since I had a child, I have worked fewer hours. I spend 2 hours a day at work pumping milk. And my husband and I alternate taking every other Friday off to reduce how much time my baby spends in daycare. I have to adjust to doing less work. Deb, Senior Research Scientist*

Deb doesn't want her baby spending too much time in daycare, so she chooses to reduce her work. But she has to get used to it implying that if she had a better option, she might not choose to work less hours. She accepts that to ensure she gives her baby the appropriate care by her own standards that she will curtail her career.

Women have to know that their children are appropriately taken care of if they are to take on a significant career commitment. And women have their own standards for what appropriate care is at each stage of their children's lives. Jen told us that a culture of leaving for family time at 5:30pm and working after bedtime was her requirement for reasonable home/work balance. Beth, the Director of Product Innovation, also told us that taking on the next level of executive commitment would cut into her home time even more—something she wasn't willing to do. All the women we talked to make these career choices with their kids in mind. No matter the solution parents find, without appropriate childcare women may not choose to take on a challenge, a promotion, or a new job—as Randi tells us.

> *My beloved manager left the company and then recruited my team to join his new company. They recruited me too. But I just couldn't get my head around how to handle the long commute, what would I do with my kids. Then at lunch, one of the guys from the team said "Get your act together, forget the commute! My wife (a stay-at-home mom) will help pick up the kids." My manager and team were at the new company, the VP was visionary, and I was stuck in a dead-end situation. So, with that support, I went. Randi, Product Manager*

Randi finally took the job she wanted once she found a childcare solution. Randi lost her dynamic team and her stimulating work, but she could not figure out how to leave the old company to join her beloved co-workers at the new company. Only with a Push from her colleague and the Support of his wife could she make the leap. With a viable childcare option in hand, Randi had choice. But until then she held back on her career aspirations.

For working parents, taking care of children is a significant challenge given the expectations of tech companies. But this challenge is less daunting if they have a partner who shares the burden. When partners share the burden of childcare and housework for real, pressure is lifted off of women. Deb told us how she and her husband alternate Friday's off to reduce daycare for their children. Here, Beth shares the equal partner relationships she has with her husband.

> *My husband and I have always worked long hours and I didn't have kids till my late 30s. I could not have done it if my husband hadn't always been supportive. I travel a lot and he deals with meals, the house and childcare responsibilities while I'm gone. Now we have a housekeeper who comes in from 3–7 every day after school. I couldn't work the hours I do if I didn't have this help. Beth, Director Product Innovation*

Beth can travel because her husband takes up the home and childcare while she is gone. So, Beth can put in the expected "good worker" hours because her husband and then a housekeeper takes on part of the expected role assigned to women. Similarly, Linda's husband shares the demands of home and family.

> *My husband picks up the house, does the grocery shopping, plans and cooks all the meals, and takes care of the kids. He quietly takes care of everything because he is committed to helping me. And it doesn't bother him that I make more money. Linda, Senior UX Manager*

Linda's husband quietly takes care of everything for their 3 kids, allowing her to focus on work. Linda tells us that at work he is a traditional "strong, silent type" as a senior manager. But at home he is right there helping, which she thinks his coworkers would be surprised to find out. So, Linda feels supported, and her husband doesn't feel threatened even though she makes more money. These supportive husbands put aside the pressure of male gender conformity and support the career aspirations of their wives. Janet and her husband also work out the home/work balance in a non-traditional way. Throughout her career Janet chose a powerful and non-traditional career—so too her approach to care for her 13-year-old daughter.

> *I made the decision very early on that I would not live a traditional woman's life like my mother who was a stay-at-home mom. So, I asked my husband to be the stay-at-home dad when he lost his job during the latest economic downturn. He is now loving it. Between us, I've always been the one with the larger salary. Janet, Director of User Research*

Like Linda, Janet's husband is not threatened by a wife making more money. When his job was lost, and since they didn't need that salary, he tried staying home with his daughter. Now he loves it. Of course, this is only one phase in their childcare balancing act. Like in many marriages, different combinations of responsibility and the help of daycare and hired helpers allow women to take on more powerful jobs, with greater time and travel commitment. And for single women, getting the daycare support mix right is essential for a demanding career, as Joy has shared. Perhaps becoming the stay-at-home or helpful husband when in your 50s after years of career success, as is the case of Janet and Linda, is easier to do than when men are young without successful established career experiences.

No matter how, families must figure out how to balance home and work responsibilities. Helpful partners and good outside support seem a must for women to taking on a demanding career. Unfortunately, companies have not figured out how to support women to have both a demanding career and a reasonable family life. Nor have they created challenging career enhancing part time jobs. In our industry, we do not make it easy, as Meg puts it so well.

> *The (Silicon) Valley is not kind to parents. When the kids were young it was really hard to find time to take care of them. Meg, Senior Engineer*

Meg tells us how hard balancing work and family life can be in tech companies and in the "hero" industries in general.[32] Companies may go to great lengths to offer a suite of options to accommodate their employees' personal lives. They offer the flexibility of reduced work hours, flex hours at work, and work from home days. Indeed, tech companies have been found to be rather generous compared to other industries in the U.S.[12] But as we have said, employees do not necessarily take advantage of these policies for fear of negative consequences.[27]

Different types of flexibility are seen in different ways. Flexible working for performance-enhancing reasons, e.g., staying longer at crunch time and making up for that another day, is likely to be rewarded. But flexibility for family reasons, e.g., reducing work hours and working "part time" is frowned upon.[21, 31] Reducing work hours leads to fewer promotions,[6] lower salaries,[13] and poorer performance evaluations.[30] So even though tech companies offer flexible policies to accommodate family obligations using these benefits is seen as an indication that a person is not really devoted to their work.[3]

The way that Jane, Deb, Janet, Beth, and all the other women deal with family and work are their way of coping with the situation. The choices that they make are individual choices. But at the same time, women's choices are never purely private and individual. Rather, they are shaped by societal expectations and the realities of how they are perceived at work when they put family before or even equal to work obligations. The choices that the women make around motherhood and work are undeniably constrained. Acting as if women are free to make choices that suit their family deny the reality of their situation. Assuming that women can indeed meet the criteria for good mothers and good workers sets unrealistic "super-mom" expectations and puts more pressure on women.[10] All of these pressures create the stress, fear, worry, and career consequences that so many of the women we interviewed express.

Social expectations for working moms create the context for legitimate worry and stress

For women in tech to respond to family demands, for them to enjoy their families, requires both Nonjudgmental Flexibility at work, supportive families, and childcare services. It would be nice if society and companies had equal expectations of men and women regarding both parenthood and work. But they don't. The ever-present daily balancing act of whether on this day family gets a little more time or work gets a little more time is simply a fact of the lives of women in tech with families. And this goes on for years—it's not about dealing with newborns.

MEN, CHILDCARE, AND CAREER

Unlike women, men simply do not have to struggle with any of these societal challenges or expectations. In general, men do not face the dilemma of being a good worker and a good parent. Being a good worker is compatible with being a good husband and father. Unlike women, when men become parents no one thinks they will become less committed to their work.[9] Whether men have children is not a consideration when seeking a new job—but women with children are greeted with more skepticism.[16] And women who are mothers are offered and receive lower salaries. But fathers receive a wage premium presumably because they are expected as good workers to support their families financially.[11] While the motherhood wage penalty has been going down, the fatherhood wage premium has increased.[14] Men with children simply do not have the career consequences that women with children face.

But single fathers are an exception. They take on both the breadwinner and caregiver roles. Then many of them reconcile these demands by letting go of career growth in order to ensure consistent availability to caregiving.[19] Single fathers do not have the option to let greater responsibility for childcare fall to their female spouse. For years, scholars and practitioners have called upon fathers in more traditional family roles to increase their contribution to domestic responsibilities. More men have taken on greater childcare roles over the years and want to be more involved in parenting and home life.[25] But unlike women, men are not expected to do these things. Men who help, who do more at home, may be championed—up to a point. Society and companies favor men working long and hard. In tech, the whole structure of work assumes that someone else is taking care of the family.[18] So, if men in tech take on too much home responsibility, they also get backlash.

If the father takes family responsibilities seriously, their stature is lowered.[17] Men who request family leave are perceived as showing too much family devotion. They are viewed as having less commitment to the organization and receive fewer rewards.[27] Becoming a stay-at-home parent is also associated with more depression symptoms for fathers, but not for mothers,[20] pointing to how stereotypical expectations influence fathers' experience. So, when men are called upon to take on more family responsibility, it is no wonder they resist—they too are reacting to social expectations and corporate pressures. They too know that if they take on too much responsibility at home their career will be derailed. Research suggests that if men used flexible work arrangements and paternity leave options it might reduce gender inequality.[33] But given social expectations of men, it is no wonder that men also do not often take a lot of paternity leave or use flexible options when it is offered.

Working dads do not face the same career consequences as moms— unless they take on real childcare responsibility

Research beyond heterosexual couples shows that LGBTQ+ families share similar experiences when balancing work and family roles. But they face additional, unique challenges. For example, most managers have little knowledge of how LGBTQ+ employees balance and distribute work and family demands between them.[24] In the workplace, perception of employees' family roles and life is still seen through the lens of the traditional two parent family where the father is the primary breadwinner and the mother is the primary caretaker. The demands of single parents and non-traditional couples are not well known or considered. For these and other styles of family life Nonjudgmental Flexibility is even more important.

Apparently too much care for home commitments is bad for any career in tech. Without very explicit pro-family values that are modeled and encouraged by managers and teams for both men and women, people will hide their childcare demands. They will continue to worry about the career impact of taking time for family. Mothers and fathers will continue to walk the tightrope that society and tech companies have created for them. For all these reasons, Nonjudgmental Flexibility is essential to keep valued workers focused on the work of the team—and to let go of their anxiety.

THE POWER OF NONJUDGMENTAL FLEXIBILITY

Like it or not, women are the societal bearers of childcare responsibility. So, understandably women are sensitive to whether or not they are doing it well, what others think about their parenting, and if they will be taken seriously as a worker. But this is also true for women doctors, lawyers, economists, businesspeople—working in any of the "hero" industries present the same challenge. As we noted in the Introduction, how companies handle home/work balance is not at the root of retaining women in tech. Women in tech still leave the tech industry more often than do women in other hero industries.

> *Nonjudgmental Flexibility combats the conflict moms face balancing societal and work expectations*

On a daily basis the conflict between being a good worker and being a good mom play out in the lives of women in tech. But we've also seen how the behavior of managers and co-workers can mitigate their stress. The Nonjudgmental Flexibility factor points to an overarching attitude that can help retain women in tech because it plays out in all of the other factors. The cohesive Dynamic Valuing team is more likely to flex to the needs of family so women can continue contributing to the team mission while doing Stimulating Work. Managers and teammates who give women The Push to take care of demands at home, and then Support them by being flexible, exert a counterforce to women's inner conflict. Then women can focus on contributing to the work without sacrificing her children. Local Role Models who model a commitment to having a home life make it possible for women to see that they too can work hard and have a family. Nonjudgmental Flexibility in the context of these positive work experiences interrupts societal messages and affirms that women's skill is wanted and valued.

So, co-workers and managers can create the conditions that disrupt the built-in conflict that women face. Organizations must recognize that societal and work culture expectations set women up for stress and the very difficult task of balancing career and children. They can create a culture and policies that fosters Nonjudgmental Flexibility. But organizational policy plays out locally, for better or worse. When teams create a culture of Nonjudgmental Flexibility, they become a powerful force for change. Managers and teams create sub-cultures that impact the expectations and actions of workers every day. The team culture can also impact the home/work balance experience of women.

It is not our focus here to imagine how to change society or corporate policy. Nor is it our goal to change internalized gender expectations. Much has evolved over the years that has allowed women to take on demanding jobs and have children—and much still needs to change. Rather, the Nonjudgmental Flexibility factor teaches us that women in tech with children can make it work when their own managers and teams make handling family a priority for everyone on the team. Then women can let go of their worry and focus on the stimulating work they want to do. Their perspectives on a diverse team can contribute to making innovative products. Nonjudgmental Flexibility at the level of the team is just good business.

Looking across the factors of The @Work Experience Framework we have covered so far, we can see that each experience builds women's skill, communicates her value, and so also builds her self-esteem These first five factors focus on how dimensions of the workplace and workplace interactions can help retain women in tech. They have implications for policy, management, and team dynamics. But to fully understand the forces at play in women's experience at work, we need to acknowledge the role of women's self-perception, self-confidence, and their internalized messages. Our last factor, Personal Power, explores the role of these inner experiences in retention in Chapter 6.

CHAPTER 6

Personal Power

When I moved to the product management job, I felt like an imposter every day. I was working with super-intelligent engineers. I was afraid that they would find out I was no good. It took me 6 months to figure out what they were even saying! I was afraid the team would dismiss me as irrelevant, so I was determined to work hard to prove myself. I told my manager how uncomfortable and full of doubt I was. She said, "Don't worry, I'm going to support you until you learn the job". And she did—she taught me the job and backed me up in meetings until I knew what I was doing. After 8 months I had a 360 review. Everyone thought I was doing a good job given where I started. My peers gave me time to grow, but I didn't! Now I know the job and the people; all that self-doubt is gone. Randi, Product Manager

Everyone has self-doubt. But many women lack confidence, feel unqualified, think that they lack skill, are imperfect, feel out of step with expectations, and are unclear about how to succeed. But with tolerance for imperfection, room to grow, positive feedback, supportive people to work with, clear expectations for success, and feeling valued and connected, women build confidence. They build their Personal Power.

In our opening story, Randi was trained in marketing and had been a successful people manager. When she was asked to step into product management as we shared in Chapter 3, she accepted the challenge. But it also threw her into self-doubt. She didn't know engineering lingo, she didn't know the product management job, she doubted her skill and that she knew what she was doing. Randi needed time to figure out the job along with training, support, and tolerance from her manager and team. She also needed to learn that her self-expectations were not in line with the expectations of others—no one expected her to be perfect immediately.

Because Randi got support and leeway from her team and manager, she had the time and space to grow into her new role. And to adjust her self-expectations. Whether women are naturally self-confident or not, the response of the organization uplifts or undermines how they feel about themselves. Women can often be hard on themselves, exacerbated by gender stereotypes. But a supportive working environment counteracts negative self-talk. As women gain experience and success, their self-confidence grows. They can learn to better manage that self-doubt. Developing Personal Power allows women to contribute their skill and creativity to their team's mission. Early career women often do not yet have a strong sense of their Personal Power—and sometimes more experienced

95% of women agree that "When I'm a valued part of the team, I'm more confident in my work." @Work Experience Survey

women do not either. To retain women in tech, managers are challenged to help women build self-confidence.

IMPOSTER SYNDROME AS A GROWTH RESPONSE

Karen's Imposter Story

When Hugh and I left Digital neither of us had run a consultancy before. We had no idea what we were doing. Back then, we had to have business cards. But even though I was the president of our company, I couldn't put President on my first card. It read consultant instead. I just didn't feel like a president! To help figure out what to do, I read books on consulting and got advice from a successful consultant in tech, my brothers in sales, and potential clients to figure out what services to sell. We set goals of how many calls to make a day and proposals to send in a week. For every phone call, Hugh sat next to me and afterward we debriefed. About 6 months later, with closed deals under our belt, I got my card changed to President.

Imposter syndrome is a buzzword used a lot to talk about women's experience in technology. Randi tells us that she felt like an imposter every day when taking on a new professional role. As she was learning the job Randi naturally had self-doubt because she did not yet know what she was doing. While she was learning her new role, she had to act as if she knew the job, before she actually did. Then, when Randi had mastered the new role, her imposter feelings went away.

When women take on a new role or a big challenge it can disrupt their usual sense of identity, competence, and confidence making them feel like frauds. Even naturally confident people can be thrown into a lack of confidence when faced with new challenges because they don't yet know what we are doing. Robin tells us she welcomes any challenge *"A blank sheet of paper is never hard for me."* But when her manager asked her to move from a developer to a product manager, she didn't just jump in.

To me, business is a secret language that I did not know. I felt like I had to go get my MBA before I could take the job. My manager didn't think I needed the degree, but I felt more prepared with it. Robin, Engineering Manager

Robin, like Randi, reacted to taking on a new unknown challenge with self-doubt and anxiety. Uncertainty about what to do drives them both to learn the job, the language of the new position, and to find people who advise. They succeed with a lot of support from their co-workers and mentors.

Taking on a new role means incorporating new attitudes, behaviors, procedures, and language into our new professional identity. This process is not unlike the development of any new role—spouse, parent, grandparent, or professional. We start by not knowing what we are doing or how to behave. We do not feel confident or competent. But we learn, try things out, succeed, and falter.

Every day we feel out of step—until we don't. We don't feel confident until we have integrated the new role requirements or skill into our unreflected-upon repertoire.

Now consider the challenge of new hires and early career women in tech. They are becoming professionals; they are taking on a new role. They work in tech where their competence is scrutinized more than their male co-workers'. So, of course, they have a lack of confidence, feel like they are not good enough, and have self-doubt. Early career women need to learn to be workers; how to act within a team, company culture, or in their job role; how to work with others; how to get permission for a vacation…and the list goes on. Every new hire we talked to—those right out of college or even those moving from one job to another—

> *Imposter feelings can come from the normal uncertainty of taking on a challenge*

expressed doubt. They are anxious, worry about their skill, and want to prove themselves right away. This anxiety is normal and natural because early career women are taking on a new role. They need to build their confidence, their professionalism, their competency, and their knowledge of how to get things done within organizations.

Uncertainty and insecurity are part of developmental processes and a normal reaction in challenging situations. It is something that people learn to overcome with experience. But classic imposter syndrome goes beyond that. It is rooted in society's stereotype of women being less able intellectually than men.[4] If women internalize this—and there is no reason to assume that women internalize gender stereotypes any less than men do—they start out with self-doubt installed by society. Then normal uncertainty in the face of challenging situations can turn into self-talk that undermines building Personal Power. Instead of attributing their achievements to their own competence, women often point to luck, charm, attractiveness, or the manipulation of other people's impressions. Women may feel that they do not deserve what they have achieved, that their skills are not the real cause of their success.

This phenomenon is magnified for women in tech where women's technical skills are continuously questioned.[6] Women internalize societal stereotypes and then the tech culture can treat them as though they might not be competent. As a result, women may develop classic imposter syndrome when facing normal challenges. To counteract stereotypes, their own negative self-talk and to build confidence, women in tech need significant support.

The Personal Power factor in our survey measures experiences related to building self-confidence which correlate with thinking of leaving a job. Looking deeper, our research shows that becoming a truly confident professional takes much longer than we might imagine. Two thirds of women aged 25–35 say that they lose confidence when others criticize them. This number declines with age, but half

> *Losing self-confidence when criticized is more pronounced when a woman is the only woman on the team*

of women aged 36–45 still agree. Losing self-confidence when criticized is particularly pronounced when a women is the only women in her team—if there are more women in the team, criticism does not affect their confidence as much. Since self-confidence is essential to really focusing on the work, these inner voices of self-doubt can keep women from bringing their full creativity to the task.

Not knowing what to do to be successful exacerbates this situation. Early career women may still be unclear about how to succeed, work in the organization, or advance. In our study, 49% of women aged 25–45 report that they don't know what to do to be successful. But the longer women have been in the same company, the more they feel they do know what they have to do to be successful. This finding also suggests that if women do not figure out how to be successful in a company, they are not likely to remain there for many years. All of these findings drove our investigation into what new hires need to be successful which produced The Team Onboarding Checklist (Chapter 7) to help managers grow new hires successfully.

The tech industry's job is to build successful professionals. Only then will we get the benefits of their diverse perspective and skill. The Personal Power factor raises up the importance of deliberately building the confidence of early career professionals and new hires of any age. This starts

Build the self-confidence of early career women by building their success

with the recommendations in The Team Onboarding Checklist. This also means deliberately disrupting and counteracting the self-doubt installed by society and the tech culture itself. But we must beware of the "fix the women" trap, asking women to behave more like men, to display confidence, and perform exceptionally all while trying to balance being assertive but still feminine enough for acceptance.[10] Instead, when managers and the team ensure that women know what to do to succeed, get support while they are learning, and are valued for their unique skill, they help build Personal Power.

SELF-CONFIDENCE AND BUILDING PERSONAL POWER

Some women say they are naturally self-confident. Some say they are naturally self-doubting. Research on women and confidence has historically shown that women have less self-confidence than men.[11] But during the second wave of the women's movement we came to recognize that traditional female gender expectations required women to undercut their own capability and defer to men. At the same time, we learned that whereas girls excelled in math when young, they eventually abandoned it when older because it was not part of the female gender role. By grade school, boys show more positive attitudes toward math than girls.[8] These differences in attitude exist even though girls receive better math grades.[12] It is still unknown when exactly gender differences in children's beliefs about math first become evident.[7] Negative attitudes toward math of course can affect whether or not girls will feel that they will be any good in fields like technology.

Research on adults in their perception of their own skill shows a confidence gap between men and women.[2] Men are often over-confident and women are sometimes under-confident when expressing their self-perception of their performance and skill. These findings occur more often in fields like tech, where assertiveness and dominance are more highly valued than qualities such as warmth and sensitivity.[13] They also occur more often in male-dominated fields when women are more likely to be the only woman on the team.[5] Being in the minority makes being a woman more noticeable. Similarly in our survey, losing confidence when criticized is directly related to the number of women in a team. Women lose confidence more easily when they are the only woman on a team. All of these findings suggest that confidence is related to the situation women find themselves in at work.

In tech, doubting women's technical ability and treating the lone woman as an outsider is certainly going to undercut her confidence. More importantly, the display of confidence is a key driver in getting and maintaining positions of power and status in the organization. Tech culture likes to think of itself as a meritocracy, that hard work begets advancement. But as we said in the Introduction, this value is a myth.[14] Instead, research finds that the display of confidence, a typical male attribute, is associated with a perception

> *In tech, confidence leads to presumed competence in men—but not for women*

of competence—confidence is mistaken for competence. On the other hand, when competence is accompanied by attributes perceived as feminine—humility, team orientation, or careful consideration of other's ideas—it is not recognized.[3] So, in tech culture women's skill is underestimated and her competent work is invisible if she does not display male-like confidence. Yet displaying too much confidence can also backfire, leaving women walking a tightrope.[17]

In navigating this situation and to maintain their tech credibility, women have adopted different strategies.[9] One such strategy is called *enhancement*: embracing the fact of being a women, making it work for you, educating others about its positive qualities, and advocating on behalf of women. Another strategy is called *assimilation*: demonstrating behavior and attitudes associated with the majority group and showing behavior that men typically show. In line with previous research, we found that late career women tend to adopt the assimilation strategy. Janet, who is in her early 50s, expresses the attitude of women who were among the first in her field. She is naturally self-confident and deliberately rejected the traditional female role. She sought and found leadership roles in tech.

> *I've always wanted to be in a leadership role. I like the power and the control. When I was growing up, I noticed that my father made all the money, had all the power, and made the rules. So, I was determined to make money, make the rules and go after leadership jobs. I've always made more money than my husband. At college, I was the first woman editor on the college newspaper—it was like an all-male club. As the editor, I could walk into the administrator's office to get information for articles. It made me feel powerful to deal with "grownups."*
> *Janet, Director of User Research*

Janet, likely influenced by the changing societal culture around her, saw the dynamics in her family and explicitly chose not to pursue the traditional female role. She has a family, but not according to traditional societal rules. She began her career in computer science at a well-known technical college as one of a few women. Like many of the women we talked to, getting a CS degree afforded her respect because of her technology skill—and the ability to speak tech. Having a CS degree bestows respect and value automatically. Janet tells us:

> I've noticed that women often don't seem to know how to talk to engineering men: direct, no nonsense, and fact based. My VP, who is a visual designer, uses flowery language, doesn't clearly ask for what she wants the engineers to do. So, she is not heard or valued by them. As a result, she cannot advocate for us. This isn't a problem for me. I have a CS degree so I can hold my own in a meeting. If you are going to work with engineers, you need to be direct, know your technical stuff, have good ideas, and say what you mean.

Janet sees herself as strong and able to hold her own with men and those with more power. She has had success after success in her work. When her work situation soured, as it did because this VP was not getting support or budget, she chose to leave without agony or doubt. Like several of the senior women managers we interviewed, Janet has confidence and knows her Personal Power. They all built it up over a lifetime of successful work.

Radiating confidence like Janet helps women's careers. Women who appear self-confident as well as caring and other-oriented are perceived as competent and as someone who meets expected performance standards.[10] Since confidence and caring are perceived as necessary for women in tech, this self-fulfilling prophecy leads to career success. But not all women see themselves as naturally self-confident—yet they too build their Personal Power through gaining experience and supportive relationships. Ellen shares her experience of stepping out of her comfort zone.

> I don't have a lot of confidence and I always worry if I'm doing a good job. I was way overdue for a promotion but I'm not comfortable asking for it. So, when my manager said there was an opening for a manager, I took it. I didn't know what I was doing but I meet with my boss every week. This gives me confidence because he gives me advice. As I did the job, everyone told me that I'm doing a good job and that makes me feel good. Any time I try to do new things and find out I'm good at it, I believe in myself more. It's not just that I worked hard to learn something. I have some real skill! Ellen, Quality Engineering Lead

Ellen worries that she has no real skill. She doesn't start with the idea that she is good and can do anything she sets her mind to. She learns that she is skilled by trying things out in an environment of support. Her manager gives her advice. He and the team tell her she is doing a good job and give her leeway to learn. Clearly, Ellen's manager and team were not influenced by bias about women's skill and the idea that only confident people are competent.[16] They see Ellen's potential to succeed even when she is not sure—and they tell her so. Trying out something new and finding she can do it builds Ellen's self-esteem and sense of competence. But trying out something new in a supportive environments makes taking that on a personal stretch goal less risky. Building Personal Power is not magic; it happens because the work environment creates the conditions for success.

Ellen also realizes that she has "some real skill." She sees her success as the result of her own actions even though she had support. Personal Power locates success within the person, not with luck or outside situations. Personal Power includes what psychologists call an "internal locus of control," knowing that our achievements are the result of our own efforts and skill. We develop an internal locus of control, like Ellen, when we take on challenges, succeed, and see that success as due to our own efforts not to luck or other outside situations. When managers and colleagues also attribute our success to our own skill, they help grow our sense of competency and confidence. Growing successful tech professionals includes helping them develop competence, skill, and so increasing confidence and an internal locus of control.[15]

> *Women grow their Personal Power when they realize they have skill—and when coworkers recognize it too*

Growing skill enhances our Personal Power, but skill is not just technical. Growing skill, women tell us, includes knowing how to deal with difficult people and situations. Knowing what to do during interpersonal conflict is a challenge that comes with time and experience. To understand, let's follow Allison's story.

Allison has been a user researcher for 10 years and is now a UX manager. She has worked at very high-powered tech companies, a startup, and now a financial institution. She has had success and failures. She has worked with supportive managers and absent or unsupportive managers. Allison has made significant contributions and she has floundered. She has had experienced people and colleagues help her and outside mentors to talk to. She is now in a company that is very valuing and supportive. She is very well respected in her organization. Here, Allison tells us how she now handles a difficult situation during a presentation.

> *I was presenting our latest research findings and a recommendation to test a new entry point to an existing feature. The audience was the VP of Product, who is known as a tough woman to present to. I started explaining the plan to understand search as an entry point to our feature. As soon as I said "search" on the first slide, the VP snapped, "Stop, I don't want to hear about search. We aren't going to address search." I took a deep breath before continuing. Then I said, "Let me finish explaining what I am talking about. After I finish let me know what you think." I explained that the test would just be an entry point to an existing feature and would not require building a new backend for search. She relented, listened, then said "Oh, you need to meet with this team and take it from there." Allison, User Research Manager*

Allison is seasoned; she knew this VP wasn't prone to listening. She knew that to do the research she thought the company needed, she needed to be heard so her approach could be approved.

But most importantly, Allison understood the VP's point of view—she had done her homework. So, she stopped, and took a breath to determine the right approach. Then Allison spoke directly into her VP's concern. She was confident in her approach. She shares:

> *I understood where the VP was coming from. Search is hard. Users expect search to be like Google and we can't deliver that quality for a small feature. I knew that many people have tried and failed to think of the best way to incorporate search into our app. But I had confidence in my idea. I believed we could learn something with the technology we already had. I knew the best way to be heard was to remain calm. I have learned over the years not to take offense at how people react. Instead, I just keep going. I did not get rocked by her interruption or irritated. I know how to deal with people when they are difficult—it's not personal.*

With experience and advice from her mentors, Allison learned that her job was more than user research. For success she needed to understand what mattered to her managers and their managers and help make it happen. Aligning one's goals with the goals of those in power is an important skill for career success, as our consulting, mentoring, and research have taught us. Both new hires and seasoned people struggle with it. If Allison wants to make things happen that matter to her, she had to learn how to frame her thinking in a way that could be heard by her VP. She had to learn to manage her response to difficult people and situations. But this was not always so. Allison continues.

> *This kind of thing has happened to me before in other jobs. When I was younger, I backed down in the moment and left the meeting enraged. I needed to speak to several coworkers before I could calm down and maybe find a way to bring the subject up again. But usually, I just gave up and avoided that person until I was forced to speak to them again. But now, I understand what I need to do. I was able to say my piece and didn't need to avoid anyone or stress about the situation endlessly.*

Growing as a professional, Allison tells us, is also about managing yourself in the face of resistance, interpersonal conflict and not getting in her own way. Unlike when she was younger, Allison no longer flies into a rage—a rage that did not serve her well. By walking away instead of expressing her rage, she at least did not do career damage. But walking away did not help her be heard by people in power or build a relationship with them. Learning to manage our own emotions is part of becoming a successful impactful professional. But Allison was not out there alone. She was already valued by her organization and supported by her manager and team. Even during this remote presentation Allison's colleagues sent supportive text messages strengthening her self-confidence.

Learning to manage our own emotions during contentious interactions grows Personal Power

Learning to manage our own emotions is part of becoming a successful, impactful professional. All employees must learn productive ways of dealing with themselves during the inevitable difficult situations that are encountered at work. People are not perfect. Interviews from the Valuing and Jerk project (Chapter 10) tell us that people don't listen, they get annoyed, can feel time constrained, are impatient, and are focused on their own goals. When people have power—men or women—handling them is more difficult. It is natural to worry about how it will impact your career. With experience, Allison has built her skill at managing people and the organization. She has credibility with the people she works with, and they give her support. Seeing herself handle this situation, Allison boosted her Personal Power and her confidence in herself as a professional.

QUIETING NEGATIVE SELF-TALK

Many of the stories we have examined reveal how women in tech—perhaps all employees—must manage their feelings and internal voices. Fear, self-doubt, and reactions of rage, hurt, feeling devalued, and more can happen in response to daily interactions at work. When our internal voices and feelings get in our way because they are playing so loudly in our minds, they can undermine our ability to focus on work or respond appropriately to others. Self-doubt, thinking we are not competent, and worrying that others will judge us are all natural feelings that come with any job. Personal Power is not just about knowing your own value; it is about acting in the face of fear and doubt. It is about changing our internal non-productive self-messages. Randi, the product manager from our opening story, worries if she is good enough, if she is doing a good job. She carries this negative self-talk into her work situations. Here Randi didn't catch an error and blames herself.

> On a recent customer satisfaction survey, 100% were unsatisfied with a particular feature. I took this personally because I'm supposed to be listening to the customer and this meant I wasn't doing a good job. I shared the result with management, and they said, "OK, this is bad." But then we fixed it. In the next survey customers were happy. My VP dropped by to thank me for fixing the problems: the customers went from 100% complaining to 100% happy. I still feel bad because I should have caught it—I put a lot of pressure on myself. But my VP is happy. I've learned that anything I miss feels a lot worse to me than it does to others. All they care about is I found the problem and I fixed it. Randi, Product Manager

Randi is stressed because she expects herself to be perfect—to never make an error—for customers to always be happy. But Randi did not let these feeling get in the way of fixing the problem. There was a problem, she fixed it, and she was complimented. She learned that her self-expectations were not reasonable; what mattered to her VP was getting the problem fixed. The responses of others helped Randi adjust her internal expectations to better match her real work context. In Allison's situation, her calm response also helped her VP take a breath and listen. People's reactions directly affect our self-esteem, feelings of value, and the strength of our Personal Power.

But what if our work context is not supportive, intolerant of failure, and unwilling to listen or communicate value? Then self-doubt and anxiety grow, especially for early career women. Amber, who is now a successful research and development manager, shares her anxiety about her competency and failure during her HCI Master's program.

> *Because I had an undergraduate CS degree, I had to be in the CS track of the HCI program. I really struggled with the required CS courses and with one in particular. My advisor told me to "show up and just do it." Neither the tutor nor the teaching assistant were helpful, and I was failing. I dropped the class. I was terrified that I would not get the 3.5 grade point average that I needed for my company to pay for the program. And I was really worried about what people would say or think because I was accepted into the Master's program as part of its diversity program. My undergraduate CS degree is from my small women's HCBU liberal arts college. There were three students in the class, and we did not have the resources to provide me with a strong CS background. Amber, Research and Development Manager*

Amber was never confident in her ability as a developer and her first few months on the job did not help her gain that confidence. She gravitated to HCI because she liked interfaces and didn't feel good about her technical skills. But her advisor insisted she stay in the CS track, which did not help her grow her sense of professional competence. Only with significant support did Amber reframe her situation and eventually pass the course.

> *I met a Black woman professor who was a well-known visual designer. She told me that my advisor was making a bigger deal about the CS course than she should, that it wasn't that important to my career. It was great to find her because I could also talk to her about being a Black woman professional in the world. I regrouped, took multiple undergraduate CS classes to get the background I missed, and took the course again. Then I had a TA who was really helpful; he was nice and willing to spend time with me every day. He dragged me through it, and I got a B. I'm glad I finished the program, but it was a horrible experience.*

Amber's anxiety over failure, fear of the possible financial consequences, and worry that she was undermining the reputation of Black students amplified her stress. Plus, without better guidance, she did not know what to do to succeed. This rendered her powerless. But in meeting a professor that she could relate to, Amber gained perspective on her situation and figured out what to do. She no longer felt alone and floundering. Her experience at school was not fun but she succeeded.

85% of women agree that "I might doubt myself, but I can regain confidence by talking to people who give me perspective."
@Work Experience Survey

Women in tech often simply do not have the internal resources, the knowledge, or the experience, to talk into their own anxiety. The job of professors and managers is to provide that perspective and step students or employees into the skills they need to tamp down self-doubt and build success after success. Co-workers also can help each other gain Personal Power by sharing their perspective and providing support. These positive work environments disrupt society's negative messages to women. Companies cannot grow women's Personal Power by willing it into being, they grow it by creating work environments committed to helping women become the best professionals they can be.

TAKING ON A PERSONAL CHALLENGE GROWS PERSONAL POWER

When women take on a work challenges and succeed, they both grow their Personal Power and realign any negative self-perceptions. When Allison chose to learn a more effective reaction to getting pushback to her ideas, she deliberately took on changing herself to become more effective. Sometimes we must reframe and modulate our reactions to become more successful. Sometimes we deliberately take on such a personal challenge. During the pandemic we were all faced with new work situations that challenged how we worked and interacted. In The Remote Work Project,[6] we found that everyone working remotely from their own home placed new demands on and opportunities for communication. In this context, multiple women chose to challenge their own style. Challenging our own style is not particular to remote working but this unusual context gives us a poignant view of the way taking on a personal challenge builds Personal Power. Let's examine some stories.

Ginny tells us how she had to change her communication style to be a more effective leader. As we said, working remotely reduces interpersonal cues.[1] Our remote research showed that across the board teams needed to change their working meetings to improve communication. They introduced more structure in how they ran their meetings and how they managed interpersonal interactions. Ginny is a seasoned worker, but she is new to product management. Ginny's manager told her that she needed to be more authoritative and less nice. Her team wanted her to control the flow of conversation in their remote meetings which had cameras off. Ginny realized that to be effective she had to change her own behavior.

> *I have a very strong aversion to interrupting and think it is very rude. When I was in face-to-face meetings, I just waggled my finger to move people along. But in my remote meetings that doesn't work so the team just talks and talks. I wasn't really taking control of the direction of the meeting and making sure things got done efficiently. I realized that if I want the floor I have to interrupt. So, I taught myself to do it. I told myself that it wasn't really interrupting, and therefore rude, but rather it was the responsibility of leadership. Now when the talk goes too long, I interrupt to say, "OK, I got it". Being nice, letting someone finish, doesn't help get the work done and it undermines my role as a leader. I also ask the developers point blank to verbally commit to a decision that seems implicit in a remote meeting. I'm not putting them on the spot, I'm ensuring they know I have heard the discussion as a commitment. I'm very proud*

---

[6] See https://www.witops.org/the-remote-work-project/ for an overview of all our remote work findings. See the Conclusion for a summary of how remote work impacts the @Work Experience Factors.

of myself for changing my leadership style. I'm getting a lot of positive feedback for controlling the meetings now—team members message me that they appreciate it. This helps me let go of the idea that interrupting is rude. Ginny, Technical Product Manager and Product Owner

Ginny knew that her aversion to interrupting others did not serve her well in a remote context. In face-to-face, co-located meetings she could wave a finger to try to direct the conversation, but this certainly did not work remotely particularly with cameras off. She realized that she had to redefine interrupting to be part of authoritative leadership, not rude behavior. If she did not control the flow of conversation and extract clear commitments, they would not get the product out the door. By reframing interrupting as appropriate leadership behavior, she changed her feelings about interrupting and was able to direct the conversation. She got positive feedback from everyone and felt proud of herself for making the change.

Every change in role or new challenge highlights aspects of self that we may need to address. Knowing how to challenge our personal style and growing ourselves professionally is part of becoming a more effective professional. Becoming aware of what we need to address in our own reactions may come from others or from a new self-awareness. Gail shares her experience.

I made an error with my project because I was reluctant to reach out for the information I needed to code my piece. I worried that I was wasting people's time by asking them to help me get the big picture of the project and thought I could figure it out. Later when my manager realized what happened he was frustrated with me. That taught me that I need to understand everything front to back. I want to be able to speak from a position of authority and stand up for my work. I struggle with self-confidence so when there is a difference of opinion of how to code things I just agree to what others decide. So, I've decided that I have to learn enough to have an opinion. I'm going to take a position as a manager even though I don't know if I will like it or be good at it. Being a manager will force me to tell others what to do, to talk with confidence to other managers, to know my projects up and down, and to take charge. Gail, Software Developer

Gail is an early career software developer working in her second job at a startup. In her previous job they didn't require her to stand up for her ideas or reach out to understand the work completely. In her first job at a larger company, they told all the new hires what the big picture was, gave them little pieces to code, and did traditional code reviews that caught errors. But now working remotely and in a startup, Gail had to take more responsibility for understanding the whole problem and getting the solution right. Because she was reluctant to ask for help—something we find to be common with new hires—she made a serious error. This was a wake-up call for Gail.

Gail's situation is exacerbated by remote working. Reaching out in a remote context is much harder than when working co-located. She no longer has the less threatening channels of co-located work like dropping in or bumping into people to start a conversation. Her manager could not drop by and look at what she was doing and provide input. Instead, Gail had to realize she needed help and formally reach out by setting up a meeting to talk the problem through—much harder than dropping in. The problem of getting help was a continuous theme in our interviews about remote working. Without informal ways of getting information all interactions occurred in planned meetings. But Gail wasn't sure if her issues were important enough to take up her manager's time. Yet as Gail told us, talking about one questions lead to another and getting a wider explanation. The remote context and Gail's reluctance to reach out got in the way of her success.

Even though this was a difficult situation for Gail, it did lead to her insights that she was not trying to "understanding everything front to back." She was not forming her own ideas about issues. Gail realizes that her natural propensity to "go along" isn't serving her well. She isn't doing her best work, which she prides herself in. She isn't the kind of professional she wants to be. So, she finds a vehicle to push herself to become the kind of professional she desires to be—taking the job as a manager. Both Gail and Ellen consciously choose to stretch themselves to increase their professionalism which in turns builds their Personal Power. Through realizations, facing their personal challenges, and growing themselves, women build their self-confidence.

Growing Personal Power is the normal trajectory of becoming a valuable professional who has pride in her work and accomplishments. But to do it requires two things. First, a series of successes at work, support from others, tolerance of growth, and clear statements of value from managers and coworkers. Women need enlightened support devoid of the pitfalls of the female stereotypic expectations. Second, women must also choose to own their personal change, reframe how they understand their behavior and their values, change their own behaviors, and change any of their negative self-talk that get in their way. If they are supported in this effort, if the organization rewards those changes, women disrupt their own and society's negative messages. When both support from co-workers and a commitment to personal change align, women build their Personal Power.

PERSONAL POWER AND RETENTION

Personal Power correlates with thinking of leaving the job. Having confidence in our own abilities allows us to confidently bring our skills, perspectives, and creativity to the work of the team. Without it, with too much self-doubt, we hesitate to fully participate. Personal Power is at once essential to contribute to the work and it is built up by success after success.

No matter if women feel confident or do not, if they are naturally confident or not, or if they are nervous because they are new to the field or a role, success breeds confidence and a sense of

Confidence in our skill is a prerequisite to full participation on the team

competence. Personal Power at work is not something we are born with; it is something that grows and develops as we become more competent professionals—and that requires support. When organizations provide the appropriate training, coaching, patience, and leeway for failure they build professional skill. When they communicate value and do not disparage skill, they build confidence. When in a supportive environment, women take on personal growth challenges and grow their own professionalism. They clear away their own barriers. Personal Power is grown as part of growing a professional.

But when support and perspective are missing there may be nothing to stop our internal voices from undermining our ability to succeed. Bonnie never got the support, perspective, or sense of connection she needed. She tells her story:

> *My father was in CS but when I was interested, he told me I would not be a good person for the field. That made me less confident in myself, but I was determined to prove him wrong. At college I majored in CS where there were only one or two women in the class. I felt self-conscious and that the professor called on the girls more than the guys. I didn't have confidence so when I was called on, I worried if I was right. And I had no friends in the class, so it made me feel more vulnerable and exposed. But I was determined to do well and in the end, I set the curve and was the best in the class. Bonnie, Research Scientist*

Bonnie is fighting an uphill battle with her self-doubt. She has the determination to prove everyone wrong; she is willing to work hard. But both at home and in school her lack of confidence and her self-consciousness about her performance rang loudly in her ears. She had no professor to give her perspective. She had no student friends to help and support her. She got the degree but did not emerge confident in her skill. This self-doubt followed Bonnie into her jobs.

> *In my first internship I really wanted to impress them, but I felt that I didn't have any support. I got hives from the stress and went to the hospital. The same kind of thing happened in other jobs when I was asked to do a hard project. I had a degree from a good school and good references. But when I got a job in a research organization at a large company, I had tremendous self-doubt. I felt that I didn't have any profound ideas to pursue, and I felt at sea in the work. My manager said it was ok but offered no guidance, and I was working alone. I know that I'm smart, but I always feel under-qualified and want to do a perfect job—so I'm stressed at work. When I got a poor performance review, it was easy to say my family needed me and leave that job. It was an easy exit. I'm happy being home now.*

Without support, without collaborating colleagues, and without a good idea of her goal and what success means, Bonnie did not see a path forward. Her Personal Power never grew; instead, her stress grew until working each day was just too much. She had a small child and a reason to stay at home, and she left the job and stopped working.

Everyone has internal voices of self-doubt some of the time. Everyone loses confidence when they make an error, or are critiqued, or feel dismissed in meetings. What counts is how the people we work with respond. Do they provide coaching, guidance, perspective, championing, or do they leave us alone at sea? Bonnie never found her footing. She never developed Personal Power. But

Bonnie also had no personal or organizational support. Personal Power is an attribute of successful professionals. For women to stay in tech they need to grow their sense of competence and confidence—and that means support.

Personal Power is grown within the context of the other factors that affect retention. The first five factors of The @Work Experience Framework point to dimensions of daily work with teams, managers, and more experienced co-workers that can help women thrive in technology companies. With a Dynamic, Valuing Team that is cohesive and tolerant of growth and home commitments, women will succeed. With managers who Push them into challenges and then ensure they are supported, women can rise to the challenge or new role. By taking on Stimulating Work and succeeding, women learn that they have skill; that they will not be held back by bias against their skill. With Local Role Models and mentors, women can get the perspective they need to regulate their internal gage of their worth. They can be helped to generate strategies for dealing with difficult people and situations—and their own doubts. Personal Power is grown in the context of a tech culture that helps women thrive.

> *Everyone is self-doubting some of the time but with supportive co-workers we develop Personal Power*

When we create organizations that produce the culture and work experience described in The @Work Experience factors, we can create an environment that interrupts bias and grows competent professional women in tech. We can compensate for the deleterious situation society and the default tech culture creates for women. Then women can build confidence and Personal Power. In a work culture of support and value, women can also address their negative thoughts and behaviors that get in their way. Organizations can deliberately create successful women professionals who will remain in the field.

Part I Conclusion

The @Work Experience Framework identifies six daily work experiences that women need to thrive. The power of a framework is that it circumscribes and focuses what organizations should examine in their efforts to retain women. Central to these experiences is the role of the team and how they experience their participation, belonging, and value. As we have said, all measures of team involvement in our survey also correlate with people "thinking of leaving their job." Retention is strongly related to how women experience daily interactions and collaborations with their team and co-workers.

Many team interactions occur in working meetings as part of core practices such as those used in design, critique, or Agile working meetings. These collaborative meetings are therefore good targets for intervention. But many interactions both in and outside of working meetings may also be experienced as valuing or devaluing—another promising area for intervention. In addition, women's experiences with co-workers reflect the overall culture of the team and the implicit rules and expectations of daily work life. Clarifying and tuning the team culture is another important intervention possibility.

The @Work Experience Framework also raises up the importance of women's experience with managers and senior leaders as role models who encourage advancement, coach, and provide challenging work. And of course, these relationships are often the best candidates to provide The Push into challenges when needed, and ensure the Support that women need for success and to build Personal Power. Tuning processes related to these relationships will improve women's experiences at work.

Taken together, we have identified a clear direction for creating interventions that can help retain women in tech. Guided by our data and the framework, we have developed a set of practical interventions, which have been used and tested with managers and product teams. In Part II, we share these interventions for you to try and tune the practices in your own organizations.

PART II

Introduction: Practical Interventions to Retain Women in Tech

In Part I we explored what women need to thrive at work. The @Work Experience Framework describes the kind of daily experiences women need and value: a tight cohesive team, stimulating work, a push into risk with support, coaching from local role models, non-judgmental flexibility to manage home and work demands, and experiences that build personal power. A recurring theme is the importance of relationships to the team and with more experienced people. Through these interactions, women receive both a sense of belonging and a commitment of support. Beneath everything is the overarching experience of being valued. If organizations can create and consistently deliver these experiences, they will both retain women in their workforce and ensure that diverse teams deliver on the promise of their creativity.

But explaining what is going on for women doesn't mean that organizations know what to do to make change. Helping organizations, managers, and teams create this daily work experience is our challenge. In the Introduction to this book, we noted that bias workshops, networking sessions, and assigned mentors haven't solved the cultural and interpersonal issues that women face. Instead, we suggest that organizations deliberately redesign everyday work practices to ensure that work life facilitates full participation for all employees, including women and diverse populations. Redesigning the way teams work and interact can help counteract individuals' biases and habits that may work against women.

In Part II we introduce tested interventions targeting areas of daily work and interactions that will create equality and help women thrive. Our goal is to give you a starting point—very concrete and practical suggestions that managers and team members can implement. We do not seek to redesign the whole organizational culture, or every practice currently used in tech work. Rather, these interventions have been guided by principles for successful process redesign and provide examples of what you can do in your own organizations. The recommendations are informed by issues that affect women but are not meant for women alone. For any change to be effective it must work for the whole team and for co-worker relationships. Only then can we achieve the business benefits of diverse teams. When women and diverse people feel that sense of value and belonging, when they know how to succeed, we set the stage for diverse teams to become the cohesive, dynamic teams that help retain women. And, we also help create teams that are more effective, creative, and produce great products.

The interventions we share here target broad areas for potential change that directly impact daily life for diverse teams: key practices, interpersonal dynamics, and team culture. All suggestions can be implemented at the local level and do not require whole organization adoption. This allows individual managers and teams to try out and make changes to their processes, iterating until it works for them. See our illustration denoting the critical intervention points we will discuss. We believe that interventions in every area are necessary to create a welcoming, valuing, and productive work environment for women and diverse teams.

Each chapter in Part II addresses one intervention. For the redesign of practices, we have selected three processes to explore in depth. We choose these practices because they are central to the everyday work practices of product teams and because our data has pointed us to their importance.

- **Chapter 7: Team Onboarding.** We learned that 50% of women are not clear about what to do to be successful. So, we developed the Team Onboarding Checklist to provide a tool for managers. This tool explicates what women and all new hires need to develop strong connections and the knowledge and skill they need to succeed. We describe our research with new hires and introduce the checklist.

- **Chapter 8: The Critique Meeting.** A maker culture is by its nature a culture of continuous critique. We found that 80% of women lose confidence when critiqued. We share techniques that work for giving feedback so that it can be heard and valued. We describe a clear, depersonalized critique process focused on collecting feedback based on accepted principles and practices in the industry, customer data, and business goals.

- **Chapter 9: Sneak Attacks on Key Processes: Scrum.** We highlight the importance of explicit practices to improve women's participation and success. Explicit practices that also ensure participation from all directly impact the overall success of diverse teams. Using the Agile practice of Scrum, we introduce The Analysis Matrix, our tool to help identify where practices, values, and expectations are implicit and so prone to bias. We provide examples of how to make small changes to an existing practice to make it both more explicit and supportive of success for all.

These redesigned practices ensure that women can participate equally in the work and get what they need to be successful. By changing the practice, we interrupt unconscious bias. But redesigning practices cannot always ensure that people interact professionally. And interactions with managers and co-workers outside of working meetings can also undercut the sense of belonging. Interpersonal dynamics is one of the most cited issues for women in tech. To explore interpersonal issues that may create a hostile work environment we created The Valuing and Jerk Project.

- **Chapter 10: The Valuing and Jerk Project.** Like everyone, women need to feel valued—unfortunately too often they do not. Women point to devaluing behavior as a key contributor to their experience of alienation. The Valuing and Jerk Project identifies which specific behaviors stimulate the feeling of value and which undermine it. Women often refer to people who exhibit these negative behaviors as jerks—hence the name of the project. We share the behaviors and the Character Posters we designed to help teams and individuals discuss how to increase valuing behaviors and become a more valuing culture.

Our last area of intervention focuses on explicating and monitoring team values. Teams can interrupt their own bias by articulating values and reflecting on how they are doing against them. By developing a shared understanding of how to work together we also ensure smooth professional collaboration. To become a resilient cohesive team, teams need to reflect and continuously improve how they work and their team culture.

- **Chapter 11: Building Resilience: The Team Manifesto and Process Checks.** The Team Manifesto helps teams define and own their values, expectations, and team culture. The procedure for defining a manifesto includes techniques to raise awareness of bias and issues important to women before the manifesto is created. Then, to hold themselves accountable to these commitments, we introduce Process Checks. Process Checks help the team monitor and tune how well they work together given their values. Through these practices, we help create cohesive, creative, and resilient teams.

All of our recommended interventions are informed by related literature, our own research, and our work with technology teams. Each recommended practice has been used and iterated with managers, teams, and individuals in the tech industry. Throughout these chapters, we discuss key principles we use to redesign practices. We end the book with a concluding chapter discussing the redesign principles we have used.

All resources we point you to are free at WITops.org. We invite you to try our suggestions and use our principles to redesign your own way of working.

CHAPTER 7

Team Onboarding

The first step in helping women thrive is making sure they are set up for success. Yet, 49% of women in our survey indicate that they do not feel guided or supported in how to be successful. We won't retain women in tech—or indeed any new hires—if they do not know what to do to be successful. More importantly, we can't help them be successful if we don't really understand what they need for success. These findings impelled us to launch research on the experience of new hires, the first process that can get women off to a successful start.

Management and human resources studies have extensively documented the importance of new employee onboarding. Employees are 58% more likely to remain with the organization after three years if they go through a structured onboarding program.[2] And companies with successful onboarding programs are also more profitable.[6] Studies reveal the pivotal role of the first 90 days of employment to build rapport with the company, management, and coworkers.[9] In the tech industry, high levels of support are one of the most important factors necessary to help new hires settle in,[11] develop a positive attitude toward the company and job, and become productive.[1] Looking at findings across 70 separate studies reveals that feeling socially accepted is a key factor in new hire success.[3] Team onboarding is a critical but often overlooked process[5] that companies can address.

> *To retain women be sure they are set up for success—team onboarding is critical*

The need for support to integrate new hires into the job reflects our findings shared in Part I on what women need to thrive. For women to be well integrated into the company they need valuing relationships, connection to the team, coaching from experienced professionals, and expressions of value from the manager and co-workers. If we want women to succeed and thrive, we need to ensure that they develop the relationships and skills they need for success. Because of bias women may not get the information and the support they need. Both the literature and our own research reveal that what new hires need to thrive and succeed is exactly what women need—especially early career women. We therefore target the onboarding process as a critical intervention point to ensure women in tech thrive. When done well, good team onboarding can help retain women in tech—and all employees.

UNDERSTANDING NEW HIRES

To gain deeper insight into the onboarding process, we conducted field research with over 50 new hires of multiple genders who had varying job roles and levels of job experience including recent graduates, very early career, and those more experienced. Participants came from large and small companies, consultancies, and startups in the United States and Canada. The data was synthesized to reveal critical themes and needs for successful onboarding. Through this process, we identified the four key timeframes and the eight building blocks that structure the Team Onboarding Checklist. Guided by the data we also identified the activities and experiences new hires needed to become integrated into their team and develop necessary skill. Following synthesis and ideation, we defined a checklist intervention as the best way to systematically encourage managers and companies to plan for and execute structured team onboarding for all new hires.

The first version of the Team Onboarding Checklist was presented at the Grace Hopper Conference in 2018[8] where it was very well received. We then iterated the checklist by working with managers of new hires who had come onboard within six months. Through a series of conversations with managers and their new hires, we defined the current structure and information in The Team Onboarding Checklist and Quick Reference Guide that we introduce in this chapter.

One overarching finding of our research is that most companies have some kind of corporate onboarding program to introduce new hires to practices at the corporate level. These programs focus on sharing the company mission, getting equipment, filling out forms for benefits, answering questions about healthcare, and sometimes skill training. They are most often intense all-day or multi-day meetings or fairs that overload the new hire with information. For most recent graduates and very early career new hires, the information and choices can be overwhelming much is forgotten. For experienced new hires, the content is repetitive of what they have heard in other company orientations, so they tune out. In both cases, new hires do not absorb or retain this information. But whereas companies deliver corporate programming, there is little guidance for onboarding new hires into the team.

Corporate onboarding does not help integrate women into the team

Although some companies are beginning to give more thought to team onboarding, few companies have a consistent, well-defined program to onboard new hires into their teams. This leaves team onboarding planning and implementation to individual managers who have little training or guidance for how to onboard new hires. Planning for and carrying out quality team onboarding takes time and effort. Especially in our knowledge-intensive tech industry, the work that managers and team members have to put in for onboarding a new hire should not be underestimated.[4] Human Resources (HR) can support them, but the responsibility for onboarding rightly falls to managers and the team who are the new hire's first relationships.

The manager and the team are responsible for onboarding into the team

The Team Onboarding Checklist was developed by understanding what all new hires need in the first few months on the job, and then organizing those recommendations into a practical simple tool to guide managers. It brings together and articulates best practices to support new hires considering the issues raised in our research by both new hires and managers in technology companies. The Team Onboarding Checklist guides managers to help them integrate new hires and ensure their success. Because what is needed is explicated in a simple checklist, it offloads some of the planning process from managers. We hope this tool will encourage managers to use these recommendations to help new hires succeed. We hope it will inspire companies to plan and structure team onboarding with the same enthusiasm they do for corporate onboarding.

Try the checklist and tune it for your company or team. You can download a free copy of the Team Onboarding Checklist and Quick Reference Guide.[7] Below, we provide an overview of key issues and the checklist.

WHO IS YOUR NEW HIRE?

Throughout Part I, we have shared the new hire experience for women as it relates to each @Work Experience factor. Here we share key considerations related to planning team onboarding based on all of our research.

THE NEW HIRE ATTITUDE

Team onboarding starts after the new hire has gone through the interviewing process and received a job offer. Every new hire is nervous about the new job. They want to know what to expect, who they will work with, what their projects will be and everything about what the new work entails. If they are moving, they want to know where they might live and information about the new location. The primary goal of all new hires is to hit the ground running and prove their worth—whether they are right out of school, in early career, or experienced. New hires want to be sure that both you and they made the right choice. New hires are looking for the following.

- **The welcome.** Am I welcomed into the group? Are the manager and the team excited that I am joining the group? Do they believe in my value and my job role from the start?

- **You know me.** Are they excited to get to know me? Do I have something in common with them? Do they take the time to share who they are with me? Will they reach out to make sure I'm included in social activities?

- **You are ready for me.** Are the manager and team ready for me to hit the ground running? Is my equipment ready? Did they think about the projects I'll work on? Have they planned the work I'll do and who I will work with? Do they understand my skill level?

- **Fast success.** Do I have an opportunity to show what I can contribute? Will they take time to help me get up to speed? Do they invite me to participate and listen to my

[7] https://www.witops.org/the-team-onboarding-checklist/.

thoughts? Do they welcome questions and new ideas? Or, are they waiting for me to prove myself?

The new hire wants to love the job they took. But they also want to know they can do it well and are making a contribution. Although they were selected for the job, they do not assume that they are automatically valued. They look to see if they are valued in the responses of their manager and co-workers to their performance.

New hires are also looking to see if you take the time to know them as a person—in their

> *New hires want to know you are excited to have their unique skills on the team*

uniqueness. New hires come from different backgrounds, different countries, and have different personal experiences, family demands, types of cognition, ethnicities, and personalities. These personal contexts help managers tune new hire activities for each individual. Team onboarding is where a sense of being known begins.

Team onboarding is where you start to plan for and flex to the individual—no matter their gender. If we want people to have a sense of belonging, a commitment to the project, the team, and the company we need to take the time to get to know them and coach them to success. Team onboarding is where connection to the team and coaching from local role models begins. Team onboarding is where work assignment, getting challenging work, and getting support in work begins. Team onboarding is where The Push and Support begins. And so, team onboarding is the first step in building personal power.

LEVELS OF EXPERIENCE

New hires also come with different levels of experience. Part of knowing the new hire and planning the right onboarding experience for them is affected by how much on-the-job experience they have in this role, in the industry, and at this type of company. As a rough guide, we have found these distinctions to be useful when thinking about the experience levels of new hires.

- **Early Career:** New graduate from any college program, first job, some experience (1–5 years in the industry).

  - Need a quick social group in the team and often still rely on advice from home.

  - Are not usually comfortable reaching out to others—set up one-on-ones for them.

- **Intern who was hired:** They don't really know the organization or group.

  - Just because they were on the team a year ago doesn't mean they know everything. People and procedures change.

  - They are still early career—treat them that way.

- **Some experience:** 6–10+ years in this industry, in the same job role, worked with similar products.

  o Give them a quick win as well as stretch goals and expect fast contribution.

  o Tell them who to reach out to; they should have more initiative.

- **Transfer:** They might know the company, but they do not know this group or team.

- **New job role in your group:** Invite them to co-create the role and the work.

Our main message is not to assume that older or more experienced people don't also need to develop the same connections as less experienced people. They also need a social group and a network—and they need coaching in exactly how things are done in this job, at this company, in this group. They may progress to independent work more quickly, but they do need the same kinds of onboarding experiences as early career new hires. Very early career new hires need a lot of guidance, to ask a lot of questions, and to find potential friends and coworkers right away. Yet, most are reluctant to reach out and bother anyone. You need to check in with them a lot. Irrespective of level of experience, new hires need guidance and support to connect and learn what to do to be successful. Being remote requires even more attention to planning the new hire experience.

REMOTE WORKING AND NEW HIRES

The Team Onboarding Checklist details how to integrate new hires whether they are co-located or remote. If people are remote, a structured team onboarding process is even more important to be sure the new hires do not feel isolated, alone, and unsupported as we have shared throughout Part I. Being co-located with your manager and Work Buddy is best for the new

Being remote makes integration into the team harder

hire so that it is easy to touch base frequently, answer questions, and provide help. But even pre-pandemic, many managers were remote, new hires were remote from their team and the best people to guide their work, and many people worked on globally distributed teams.[4] Then the pandemic created a remote work situation where everyone was remote from each other.

Many new hires during the pandemic went through a fully remote onboarding process, which at least in our data did not go far enough to ensure that new hires were integrated and guided to success. Both our new hire and remote work research helped us understand the needs of remote workers. A study of new hire onboarding of software developers at Microsoft during the pandemic dovetails with our own research and our recommendations in the checklist. Their findings show that the biggest challenge new hires face is building a strong connection with their team.[10]

Being remote for onboarding exacerbates the new hire's situation—it is simply harder to form connections, get coaching, develop necessary skill, learn the team culture, and generally come up to speed. And in a remote context, informal interactions and information exchange are nearly nonexistent; all interactions with the manager, buddies, and other co-workers must be set up explicitly. Even more experienced people may be hesitant to bother people in a remote context. The tips for onboarding remote new hires are the best-in-class recommendations from our research.

A remote work situation heightens the necessity to plan and to be explicit about what is needed to bring on new hires. The practices and recommendations of The Team Onboarding Checklist are even more important when working remotely to ensure that new hires are integrated into the work and life of the team.

Tips: Onboarding Remote Workers

If you are onboarding a new hire who is working remotely, here are ideas for managers to make a connection right away. This presumes that some in-person interaction is possible.

- Share the best way to contact you and encourage reaching out.
- If possible, travel to the new hire their first week and get them started if possible.
  - If the new hire is remote to you and the team, have the new hire travel to you.
  - If corporate onboarding is in-person and near the core team, travel to meet the new hire there for the first week.
  - If you can't get together early on, talk by video every day.
- Find local go-to people for answers of any kind.
  - If the new hire is in another country or challenging time zone, connect them with people in their location or time zone so they can call/meet/video during work.
- Set up frequent contact to establish a habit of conversation and help.
  - Face-to-face video or in-person meetings.
  - Keep the video on for selected hours where everyone on the team or buddies can work side-by-side and be available for quick check-ins.
  - Keep video on in meetings especially for the first few weeks.
- Drop in by phone, text, messaging, or other channels to spot check how the new hire is doing.
  - Use group channels for the team to drop in and help also if they are remote to new hire.
- Ask the new hire how the communication plan is working and adjust it.

OVERVIEW OF THE TEAM ONBOARDING CHECKLIST

If we want women in tech (and any new hire) to stay, we must communicate that they are welcome. We must show them that we will help them connect to the team and the company. We must help them succeed right from the beginning of the relationship. The behaviors needed for successful team onboarding are not mysterious, but they are often not clearly articulated or performed. Onboarding a new hire into the team takes deliberate planning and relationship building. And it starts before the First Day.

The Team Onboarding Checklist is structured into four important timeframes and eight building blocks defining key relationships and information that new hires need to succeed and thrive. The timeframes and building blocks emerged as important contexts and issues for new hires that we interviewed. The recommendations in the checklist reflect what worked and didn't work for new hires in the onboarding process. The structure of the checklist itself, its level of detail, and its content were defined and iterated both from the data and with managers. See the Quick Reference Guide for an overview.

1 • BEFORE FIRST DAY	
CONNECTION	**SUCCESS**
MANAGER	**TOOLS & INFO**
• Communicate your excitement • Give overview of what to expect • Answer questions • Get a personal statement	• Communicate your excitement • Give overview of what to expect • Answer questions • Get a personal statement
BUDDIES	**TEAM CULTURE**
• Pick the Work & Support Buddies • Ask them to reach out to NH before	• Share real culture: hours, work at home, dress, group activities, fun • Share home/work in practice: work-at-home day, sick kids...
TEAM	**JOB ROLE**
• Give the team an overview of the NH • Ask them to reach out to NH before	• Share excitement & expectations for role • Set up 1-1's with key work collaborators
NETWORK	**PROJECT**
• Identify possible friends at work • Share relevant interest groups	• Share first project & collaborators • Plan the next 3 months of projects

2 • FIRST DAYS

CONNECTION

MANAGER
- Day 1: Greet & spend time together
- Give overview of the business, work area & your management style
- Invite & answer all question

BUDDIES
- Work buddy introduces the job,
- techniques, & team
- Support buddies share their perspective

TEAM
- Hold a fun welcome event; share NH strengths & fun bio's
- 1-1s with key co-workers; not just drop-in's

NETWORK
- Connect NH to potential work friends
- Share goals & expectations of key managers

SUCCESS

TOOLS & INFO
- Day 1: all equipment, tools & access tp information working.
  IT Buddy is point person.
- Start reviewing Team/Job links & deck

TEAM CULTURE
- Help NH learn the team culture & daily work-life activities
- Share best way to communicate with co-workers

JOB ROLE
- Share work techniques & expectations
- Meet with collaborators doing the same job role

PROJECT
- Start NH on first project with Work Buddy
- Share work plan for the first weeks

3 • FIRST WEEKS

CONNECTION

MANAGER
- Check in often, beyond 1-1's
- Pay attention to NH overall ad-justment

BUDDIES
- Work Buddy partners on first project; guides & gives feedback
- All buddies check in & answer questions

TEAM
- NH continues 1-1's with co-workers until all met
- NH participates in work meetings; invite their ideas

NETWORK
- NH meets influencers to understand their priorities
- Co-workers invite NH to interesting events

SUCCESS

TOOLS & INFO
- Ensure all tools & information can really be accessed
- Company Buddy: Introduce Company Document & key physical places

TEAM CULTURE
- Explain/train NH in team & role work techniques
- Explain values & appropriate behavior for meetings

JOB ROLE
- NH shadows & participates to learn job role
- Explain how NH job fits with other collaborating roles
- Explain role boundaries & watch for conflict

PROJECT
- Monitor NH work quality & success
- Give lots of feedback & examples of quality
- Share work plan for the first 3 month

4 • LAUNCH	
CONNECTION	**SUCCESS**
MANAGER • Monitor if NH is launched • Give NH challenges • Explain path to career success	**TOOLS & INFO** • Ask NH to help others with tools • Ensure NH understands important corporate rules
BUDDIES • Buddies remain go-to support people • Look for a Career Buddy to guide NH's professional growth	**TEAM CULTURE** • Monitor team culture to ensure the team is working well after introducing new people
TEAM • Ensure NH fully participates in the team's work; that they are heard & valued	**JOB ROLE** • Monitor progress. Is NH successfully delivering, collaborating & influencing? • Coach success for NH's level of experience
NETWORK • Showcase NH success to influencers • Help NH key career influencers • Ensure NH is connected socially	**PROJECT** • Plan work & participation challenges appropriate to NH skill level • Check if NH is happy with projects, adjust accordingly

THE FOUR KEY TIME SEGMENTS IN ONBOARDING

Activities in the checklist are organized by the four time segments that represent the important phases in the onboarding process. Each time segment has different goals and activities that will help the new hire acclimate and become productive. These include:

> **Before First Day** (after accepting the offer). This is the most neglected timeframe. Once the job is accepted, the new hire is anxious to know what will happen on the first day. They want to know what awaits them—and if there is anything that they can do to prepare. No matter the company policies, the manager can always do something to help prepare and welcome the new hire before day one.

> **First Days** (1–3). The very first day on the job may be consumed by corporate onboarding. But within those constraints, the manager can meet the new hire and

connect them to co-workers, get equipment up and running, and introduce the first project. First Days are where you show the new hire you are prepared for them.

First Weeks. The First Weeks make or break the new hire's sense of competency and belonging. They collaborate with co-workers, do projects, and participate in meetings. They look to see if you will take time to bring them up to speed, if you think that you made the right decision to hire them, and if this job is a long-term fit for them. And they are looking for friends at work.

Launch (8–16 weeks, depending on the new hire). New hires are still learning months into a job. But they also want lots of feedback and start thinking about promotion. Managers look to see if the new hire is really launched, works well on their own and with others, and speaks with confidence. Do they deliver on time with quality, leverage learnings from previous projects, participate in meetings confidently, and talk in one-on-ones, to a group, and with an influential person? Whatever the answer, the manager must tune the planned activities.

Each timeframe represents changes in the new hire's self-perception, expectations, and what is needed to step them into success. Each timeframe requires different activities and goals for growing this new team member. Within each timeframe, we characterize the planning and activities recommended to address the phase organized by the eight building blocks.

THE EIGHT BUILDING BLOCKS OF TEAM ONBOARDING

The eight building blocks represent key people and information that new hires need for success. Four building blocks help the new hire connect to the key relationships they need to develop to become part of the team and to learn the company. Four building blocks support planning for the new hire's success at work. These include the information, tools, and practices critical to delivering value. As we said, although these building blocks emerged as critical for all new hires, you can see these are also the relationships, knowledge, and support that we called out in Part I as necessary for women to thrive. We introduce the building blocks below.

CONNECTION BUILDING BLOCKS

Connection building blocks describe the relationships managers must help the new hire build.

The Manager is the Linchpin of Onboarding

The manager executes and oversees all planning for new hires and curates the new hire's connections with buddies, team, and a larger network of influential people. The manager also plans the new hire's first projects and ensures that they get the information they need for quick success.

Few managers are trained as managers, let alone guided in best practices for onboarding. Busy managers, those expected to do individual contributor work as well as manage, those onboarding multiple new hires at once, or those with many reports may feel they do not have the time to plan and closely guide new hires.

Moreover, the gap between onboarding at the corporate level and onboarding into the team leaves holes in the hand-off. The company may, for example, have a relocation service but a relocating new hire still wants someone they trust to

> **CONNECTION BUILDING BLOCKS**
>
> **MANAGER:** Responsible for planning and overseeing the onboarding experience; the first line of connection critical to success, particularly for early career
>
> **BUDDIES:** The set of people who will partner with and guide the new hire in doing the work (Work Buddy), getting tools and tech working (IT Buddy), providing experienced perspective and feedback (Experienced Buddy), and understanding the company and being an employee (Company Buddy).
>
> **TEAM:** The people who will work directly with the new hire within and outside of the formal team; the source of value and knowledge
>
> **NETWORK:** First the new hire's social connections, then other collaborators, influencers, and stakeholders within the company; critical for career development

talk to about the city and where to live. New graduates may need help filling out forms, knowing what equipment to select, or picking health insurance. They naturally reach out to their manager before their first day.

The manager is the new hire's first relationship and so the first stop for help. The onboarding relationship is the beginning of the supportive relationship that all new hires need.

Buddies are the Sidekicks

New hires who do well form a key set of formal and informal buddies who guide and provide information. In Part I, we discussed the vital role of buddies, partners, mentors, and guides for women to thrive. But it turns out that all new hires need to develop these same relationships. The buddies we define in the checklist codify these critical first relationships. They are the new hire's go-to people for questions, no matter how dumb. We recommend that you assign these key buddies explicitly. By reaching out, supporting, and partnering, these buddies show the new hire that they are valued and belong—right from the start.

Work Buddy. This buddy is the new hire's potential first friend at work. Select someone with more experience doing the same job, preferably near to the new hire's age and life circumstances. The Work Buddy partners with the new hire for their first projects. They introduce the new hire to all aspects of the work and life of the team.

Experienced Buddy. This buddy is the first local role model, experienced coach, and the source of big-picture information about the work and job. If the manager does not take this role, find the right senior person. It works best when these buddies are local and have the same job role.

IT Buddy. This buddy is someone on the team who understands the tools and systems used by the team. This is not corporate IT that might help as part of corporate onboarding. The IT Buddy is that go-to person in the groups that everyone asks for help when they are stuck. The IT buddy ensures that the new hire's tools and systems are set up to access everything they need for their work. Too often organizations confuse the Work Buddy and IT Buddy. Just because someone can help with tools and systems doesn't mean they can or want to partner and connect with the new hire—as some of our managers found out the hard way.

The Company Buddy. This buddy is anyone who really knows how to navigate the company. Sometimes it is an administrator in the group or sometimes a senior person takes on the role. The Company Buddy explains the way things work and where to find all the information the new hire needs to navigate the corporation and department. They also go out of their way to welcome and help the new hire become familiar with how to get things done in the organization.

We cannot overemphasize the importance of these buddy relationships for new hire success, and also for the continuing success of early-career professionals. Buddy relationships work well when they are an expected part of the team culture and being a buddy is valued from a career perspective. If being a supportive buddy becomes "team housework"—a devalued activity—new hires may not receive the support they need. If companies depend on volunteers to play these roles—as in the case of the Company Buddies we talked to—they risk losing the function when these good Samaritans leave or are simply too busy. We encourage companies to recognize and plan for all buddy roles.

The Team is the Source of Belonging

The team is composed of the new hire's co-workers who they work with on a daily or frequent basis. For women, the sense of belonging and valuing by their team is central to thriving at work—so too with all new hires. Integration into the team is at the center of team cohesion and successful teamwork and collaboration. New hires must be well integrated into their teams.

New hires may belong to several teams, for example, new hires are in a team of those cross-functional co-workers they must work with to get a project done. A typical product team will include product managers, developers, designers, researchers, content designers, and more. In addition, those who report to the same manager or who do the same job may also feel like a team or group that new hires must get to know and connect to. New hires need to develop relationships with whoever they will work with frequently —these are their key co-workers.

Successful integration into the team is critical for building team cohesion

But too often team onboarding looks like a quick tour of the people in the hallway to say hello. Best-in-class managers set up one-on-one meetings with all the critical relationships the new hire must develop. They also task the Work Buddy with bringing the new hire to all important team meetings, events, or social gatherings. They tell the team to reach out through LinkedIn or other communication channels both Before First Day and in the First Days. In other words, they make sure the relationships happen.

Managers who give a new hire a list of people to connect with on their own may be disappointed. Early career and many new hires are not comfortable reaching out or asking for time from a stranger, particularly a busy co-worker. And as we said, reaching out to a remote coworker is even harder. A lone woman on a team may be especially uncomfortable reaching out. The experience of The Welcome happens when the team reaches out to the new hire to make relationships. Successful integration happens when all of the new hire's co-workers, irrespective of department, see themselves as part of the onboarding process—and reach out.

The Network is the Community of Influence

From the start, new hires need to create a social network—they are looking for friends at work. But in addition, networking with people in the company is essential to advancement and getting interesting work, as women and new hires have told us. Surprisingly, some managers we interviewed did not know who might be in the new hire's network or why building a network in early career is important. They also did not realize that new hires begin to wonder about where they stand and promotion within a few weeks.

Managers need to help the new hire develop a network of influencers. Key influencers in the company include anyone more senior inside or outside the team who will influence or evaluate the new hire's work. These may be influencers in their management chain, members of collaborating organizations, and senior professionals with the same job type. Best-in-class managers are able to identify these influencers, know what matters to them, and help new hires understand how their work contributes to these goals. For key relationships, managers can set up getting-to-know-you meetings. By First Weeks and certainly Launch, managers can encourage new hires to speak up at larger group meetings and create opportunities for the new hire to showcase their work. Expecting the new hire to know who to reach out to and how to interact with influencers in a new company is not reasonable unless they have a lot of experience and stature.

Identify key relationships and set up first meetings for the new hire—don't expect them to do it

Managers who get to know the new hire as a person can also help connect them to professional and special interest groups in the company. These groups are a source of both friends and local role models. New hires value being invited by a co-worker to come along to these group experiences; when invited they often try it out to meet new people and to connect to that co-worker.

Telling new hires—even more experienced new hires—about a professional group or one related to a shared interest may not be enough. Invitations from co-workers to go together are the best ways to help new hires connect to a larger network. The people they meet through these group experiences may become a source of information and shared experience needed for advancement. We encourage managers and the team to help new hires develop their networks.

The Connection Building Blocks ensure that all new hires build the relationships they need to feel connected, to feel known, to advance their career, and to get the work done with quality. These relationships ensure that new hires are integrated into the company and their teams. The manager's job is to plan for and oversee these connection activities to help new hires form important relationships.

SUCCESS BUILDING BLOCKS

SUCCESS BUILDING BLOCKS
TOOLS AND INFORMATION: Access to all tools relevant to new hire's job role, typical to the team, domain information, past project examples – with helpers to find & interpret how and when to use it.
TEAM CULTURE: The team's way of working and socializing, rules of engagements, habits, and expectations beyond corporate policy.
JOB ROLE: Clear communication of the value of the role. Clear role expectations, responsibilities, practices, and typical collaborators.
PROJECT: The project or work that the new hire is assigned to with clear expectations of deliverables, timeframe, and success criteria.

Success Building Blocks ensure that new hires get this knowledge. But too often what new hires need to know is unconscious, unarticulated, or hidden away in complex internal websites. Success Building Blocks make this knowledge explicit and identify people to explain it. (See the description of the success building blocks.)

Tools and Information

All good corporate onboarding programs help new hires get up and running on their computer, mobile devices, digital tools for the job, and access to team information. The goal is up and running on Day 1. When everything is not immediately available, new hires are frustrated and feel the company is not "ready for me."

Beyond these initial tools, new hires also need information to successfully navigate the company, their division, and their team. For example, they must access information on company benefits, travel, rules, mission, presentation templates, etc. They need to find physical places: printers, fire doors, mothers' room, and lunchroom. Larger companies may have many internal sites for company and department information. But sites are often numerous, overwhelming, of unclear relevance, and hard to use. Smaller companies may not have any information codified, let alone readily accessible or organized.

Through our interviews, we found good Samaritans in some companies who create custom documents and slideshows introducing relevant information. They also meet with the new hire for a tour of relevant places and to walk them through the information. These good Samaritans inspired the Company Buddy role. We recommend formalizing the Company Buddy role. The Company Buddy, along with the support information they create, becomes part of the new hire's experience of The Welcome.

To get their sea legs, new hires need to figure out how to "be an employee here." The challenge is greater for recent graduates who are trying to figure out how to be an employee for the first time. We recommend creating an Introductory Company Document outlined in the appendix of the Team Onboarding Checklist document.[7]

> *Organize the information new hire's need to understand working in this company—and walk them through it*

Once created, managers or HR people assigned to a department can use and tailor it for their departments or new hire needs.

Team Culture and Practices

New hires need to be introduced to the team's work, culture, and practices. New hires look at everyone's behavior and ask questions to try to figure out expectations and team culture. A fun example was shared by a new graduate developer: *I saw that everyone wore Dockers on the team and so I bought a pair.* Team culture is more than clothes, everyone eating together at lunch, and standard social time. It's also how the team runs itself. New hires need to know the days and hours everyone is in the office (or if new hires are remote, when people are available). Teams have standard ways of doing things like the weekly critique or group meetings, processes, and slideshow templates including how much deviation from the template is accepted. Teams also have their typical communication and work tools that everyone is expected to use. New hires need to know these expected practices so that they can fit into team life. They do not want to violate traditions and expectations or fail to show up at the right time to the right gatherings. Telling them directly takes away the guesswork.

Unfortunately, managers don't always have a clear idea of team culture and practices. Several managers we worked with became inspired to brainstorm with their team to figure it out. Managers and teams cannot be sure their culture is inviting if they cannot even describe it. And, if the culture is very male-oriented, women may be alienated and not want to participate. Work with your team to articulate your team culture and practices, then you can choose what you want it to be.

> *Create a team document explaining the traditions, expectations, and practices of your team*

Best-in-class managers create a document or slideshow to introduce standard practices of the team to the new hire. This often includes team-level links to information, descriptions of team culture, and an introduction to relevant practices, tools, deliverable examples, and links specific to the new hire's role and project. Create this document and iterate it over time. Use it to make how you work as a team explicit. An overview of key contents for a Team Document can be found in the appendix of the Team Onboarding Checklist document.[7]

Job Role

Each new hire has a job role: developer, product manager, user researcher, designer, quality manager, compliance expert, trainer, etc. Each role comes with different responsibilities, expectations, decisions they are responsible for, and implicitly or explicitly their power to influence others. But how that role is valued, the exact expectations, and ways of working typical of that role are different in each company and even within each group. The new hire must learn how to be successful in their role at this company.

To help new hires acclimate quickly, the manager, or an Experienced Buddy in the same job role, must explain the role and the details of the practices used in the company and for this job. Early career new hires have degrees related to a specific job role, developer, or designer for example. But a university doesn't prepare the new hire for the realities of expectations for the role at this company.

Clarify and communicate the expectations and practices of this job in this department

The experiences at another company may not map or carry over to this one. As some women we interviewed discovered the hard way, how things are done at one company, how they are evaluated, how much influence they can use, and how people with that role are treated is different from company to company.

Best-in-class managers champion both the new hire and their role. They help them learn the level of influence they can wield appropriate for their level of experience. Good managers also provide air cover and protect the boundaries of the new hire's job. For example, one development manager stepped in when a product manager kept asking a female new hire developer to build a user interface instead of asking a designer. A new hire doesn't know what is expected—the manager needs to communicate expectations explicitly and protect that work.

The job role building block makes the unspoken expectations and realities of the job clear to the new hire; they encourage a conversation about the job. This clarity not only helps new hires. It also helps women know what to do to be successful right from the start.

Project

All new hires want real work right away so they can have a quick win. Success on a project allows them to show their value and that they are the right person for the job. Being ready for the new hire means that you have planned their projects and colleagues to work with. All new hires, no matter the experience level, want a Work Buddy for their first projects. New hires know that they are learning your systems, architectures, design patterns, and tools. They want to do that while producing real deliverables.

Best-in-class managers create and share their plan for a first project in the Before and in First Days timeframes. Then they share a 30-60-90-day plan. As managers get to know the new hire, the plan can be tuned. Good project planning focuses on how to step the new hire into the work. As the new hire gains increasing skill, managers increase the project challenge in a stepwise fashion. For example, a new female developer began a 1–2-week project with her Work Buddy that taught her the system, coding expectations, tools, and the process of code review—but also delivered a real piece of code. Then the pair moved on to something harder that took longer with other learning challenges. Similarly, designers shadowed their Work Buddy to learn how they work and then co-created a simple design that was critiqued by the group. Experienced new hires are also stepped in, but they and their Work Buddy start with a stretch goal that will teach the team's systems, tools, and practices.

> *Create a stepwise plan of increasing challenge for the new hire's first projects—with a Work Buddy*

Planning increasingly challenging projects is the best way to be sure the new hire becomes excited about the job. Although they will do it, new hires complain if managers always ask them to do uninteresting work even if it is "available or must be done." When one team discovered that the new hire was good at bug fixing, they started bringing him all the bugs. The manager stepped in to ensure the bug work was passed around.

Planning challenging projects is also an act of valuing for the new hire; assigning a challenging project is seen as an act of confidence in their skill. Planning increasingly challenging work for all new hires also creates a habit of planning for challenges for all employees—a habit that we hope will help women get stimulating work. Planning challenging work is part of planning for retention.

The Success Building Blocks ensure that new hires get the information they need and develop the skill necessary to be successful. The activities identified in the Success Building Blocks of the checklist ensure that what people need to know to be successful is articulated, well defined, and easy to share with the new hire. The checklist highlights what managers and the team can do to help new hires come up to speed at the company, on their team, for their role, and to contribute to projects. The Success Building Blocks directly address the problem that women don't know what to do to be successful.

STRUCTURE OF THE TEAM ONBOARDING CHECKLIST

The Team Onboarding Checklist is designed to include the information managers need to create their own plan for a new hire. It is structured into the four timeframes providing recommended behaviors and activities in each timeframe relevant to each of the Connection and Success Building Blocks. Breaking the activities down in this way helps managers think about what matters for integrating new hires. And it ensures that the task of planning for new hires doesn't become too overwhelming.

The elements of the checklist can be integrated into other onboarding requirements guidelines or checklists to produce an integrated document to guide managers. We encourage managers, groups, and divisions to redesign their own plans inspired by our checklist. The overarching goal of the Team Onboarding Checklist intervention is to raise awareness of what is needed for new hires to succeed and provide an easy-to-use tool. Since the checklist also ensures that women get the information and support they need from the beginning, it goes a long way to creating the work conditions needed for retention.

We have included the Before section of the checklist to give you an idea of how the checklist is structured and the content it provides. (See the Before section of the full checklist.) Within the checklist, each timeframe presents suggested behaviors organized by the Connection and Success Building Blocks.

BEFORE: CONNECTION		
MANAGER	**DETAILED ACTIVITIES**	**YOUR PLAN**
Communicate your excitement	Reach out to new hire in a face-to-face meeting, email, or by phone. Communicate that you and everyone is excited they are coming. Give them your contact information.	The HR/Manager Gap
Give an overview of what to expect	Provide a schedule and high-level overview of the first few days' activities. Include a brief introduction to their first project and key people they'll meet.	
Answer questions	Answer questions about paperwork, relocation, vacation, or any other issue the new hire raises. Provide help.	
Get a personal statement	Ask the new hire to write a short paragraph with a bio, key interests, and a picture to share with others.	
BUDDIES	**DETAILED ACTIVITIES**	**YOUR PLAN**
Pick the Work Buddy	Pick a Work Buddy who does the same work, is co-located, and has a few years' more experience than the new hire. Communicate your expectations and how you will monitor their success.	Learn about buddies
Identify other support buddies	Identify a set of buddies to help the new hire increase skill (Experienced Buddy), introduce the organization and corporate websites (Company Buddy), and a technical helper (IT Buddy). These are in your team or group.	
Ask buddies to reach out	Ask all buddies to reach out by email, LinkedIn, Slack, etc. to communicate their excitement, explain how they can help, and share something personal.	
TEAM	**DETAILED ACTIVITIES**	**YOUR PLAN**
Give the team an overview of the new hire	Share new hire's resume, skills, portfolio, LinkedIn, personal email (if the new hire agrees) to get the team excited. If you asked new hire for a personal statement, share this.	
Ask the team to reach out Before	New hire's key co-workers reach out through email, LinkedIn, or an acceptable channel to communicate excitement and any common interest or experience.	
NETWORK	**DETAILED ACTIVITIES**	**YOUR PLAN**
Identify possible friends at work	Help the new hire create a social community among the people they will work with. Identify others in their same age cohort, especially with the same job.	Who is in the new hire's social community?
Share relevant interest groups	In your first contacts, let the new hire know about any company or local clubs or Slack groups they would be interested in based on their personal statement. Include relevant centers of excellence for their role.	

BEFORE: SUCCESS		
TOOLS & INFO	**DETAILED ACTIVITIES**	**YOUR PLAN**
Be ready	Ensure that all computer equipment, phone, office equipment, etc. the NH will need is ordered to be ready for Day 1. Find a desk for the NH to work near the team.	Prepare introductory documents.
Give NH choices for equipment and healthcare before they come	Let NH fill out forms or at least learn options for healthcare, benefits, etc. before official start date. Give them time to choose and discuss. Use a Hub, not email and fax to sign and select.	
Give NH a jump-start to understand company and job	Send links, marketing descriptions, articles, process descriptions or other non-proprietary information about the company, product, or domain they will be working in. This is not required work; it's context information.	
TEAM CULTURE	**DETAILED ACTIVITIES**	**YOUR PLAN**
Share real culture of the team beyond corporate policy	Real work practices: typical hours people are in the office vs. work from home, usual dress, group lunches, fun stuff... Real home/work balance: work at home days, vacation, sick kids, overall home/personal values of the group.	
JOB ROLE	**DETAILED ACTIVITIES**	**YOUR PLAN**
Express excitement & value of NH role	Provide a specific description of their role and your expectations (even if you discussed it in the interview). If the job is a new role, talk about how the job will designed together.	
Set up meetings with key work collaborators	Set up 1-1's over the 4-6 weeks with key work partners in your team and collaborating teams. Include people who have done the NH's job role. Don't ask early-career NHs to set up the meetings themselves. Doing the set-up for any NH is valued.	
PROJECT	**DETAILED ACTIVITIES**	**YOUR PLAN**
Get NH excited about their first project	Share a high-level description of the first project including the schedule, who NH will partner with (Work Buddy), and how the project will help them learn.	
Plan NH projects for 90 days	Plan out NH projects with co-workers for the first 3 months. Give NH a quick success. 30-60-90 day plans work. Be ready to share the plan in the first week.	

Looking at the checklist structure, the left column is a quick reference to the recommended action. The middle section provides more detail of the behaviors. Through discussions with managers, we learned that managers benefited from increased detail about the activity. The right column provides a place for managers to write a plan. The links in the planning section reference issues to consider for that timeframe and building block. These issues are available in the Appendix of the checklist organized by timeframe.

To encourage managers to continuously think about what is needed for onboarding a new hire, we created the Quick Reference Guide shared above. It is available as a printable sheet or a stand-up card for your desk. You can download the full free checklist and quick reference on our website. [8]

THE TEAM ONBOARDING CHECKLIST FOR RETENTION

The Team Onboarding Checklist provides a structure and a process to help improve planning for onboarding. The contents of the checklist come from our research data from new hires and their managers of all genders and backgrounds. Helping managers focus on key success factors for all new hires through a simple tool like a checklist ensures that all new hires, including women, will be more likely to receive the same onboarding experience. The Checklist helps managers plan for what is needed for success for everyone.

The Team Onboarding Checklist also encourages action, reflection, and explicitly defining what new hires need to be successful. By articulating the dimensions of work life that help new hires succeed, we encourage managers to think about what might be missing from their current approach. For example, until we proposed the idea of contacting new hires during the Before time segment, managers did not realize how much it meant to new hires and how restricted managers felt by their company policies. Through our conversations, we explored what managers could do within those constraints. Similar conversations about team culture, stakeholders, and buddies helped managers and teams find ways to do a better job. Simply by understanding the new hire's experience and the best practices we encourage reflection and nudge change.

A structured articulated process like a checklist is one of the best ways to ensure that women receive the same treatment as men—that *all* new hires are guided to team integration and success. Making practices and expectations explicit ensures that everyone knows what to do to fit into the team culture and be successful. By asking managers and teams to provide the suggested information, explain expected practices, and help new hires create needed relationships, we nudge organizations to make explicit what was implicit for work success. By being explicit about culture and work expectations, for example, managers we worked with were more likely to become aware of things that might not really work for

A structured team onboarding helps women know what to do to succeed

women. Structured explicit practices like the checklist can interrupt unconscious bias that may disadvantage women. By knowing what works for all new hires, people of good will are less likely to deliberately treat women differently than men.

If we want women to thrive, if we want to make sure everyone is treated the same from the start, The Team Onboarding Checklist provides managers and organizations a structured process to achieve those goals. The Team Onboarding Checklist is the first step and a primary stopgap to bleeding women out of your company. It is a tool for retention for all employees.

---

[8]   https://www.witops.org/the-team-onboarding-checklist/.

Now we turn to another important process for women, giving and receiving critique on a work product.

CHAPTER 8

The Critique Meeting

Technology is a maker industry. Diverse teams design and develop products, websites, apps, systems, services, cars, refrigerators, medical equipment, and more. Like any maker industry, making comes with getting continuous feedback or critique.[7] But 80% of the women we surveyed lose confidence when they are criticized.

The practice of critique is both how work is improved, and how organizations improve the development and design skills of their workers. The Critique Meeting is not an individual evaluation or performance review. It is not meant to give a grade or to determine if presenters will get a bonus promotion. Rather, The Critique Meeting is focused on how to improve the work product

Become an effective feedback culture

being presented. It is also how teams coach early career professionals to improve their practices and align them with organizational expectations. Critique facilitates experiential learning and reflective practices, thereby creating the context for ongoing learning for both the person and the team.[5] When done well, it improves the product and the individuals.

We have targeted The Critique Meeting as an intervention point because it is a key process for technology teams to do well. No great work product is created without critique. No team member grows professionally without good feedback on their work. Being able to give and take critique well and learn from it is essential in tech. Yet, women and other diverse populations often do not feel heard or that their work is valued. So, interventions in The Critique Meeting are a good starting place to ensure teams get feedback from all voices in the room and that feedback does not come across as judgment of one's personal value. To retain women in tech, we must address how to become a well-functioning feedback culture.

For a typical product team, work is assigned to different individuals or sub-team members. Each person or sub-team is tasked with doing their part of the work to complete the whole product. Depending on the work product, for example, team members may research, design, test, or develop their part. But all parts must come together into a whole that works for the customer and the business; the parts must come together coherently into a high-quality, excellent product. Critique ensures that multiple perspectives come to bear on the creation of each part in light of

Feedback from multiple perspectives increases product excellence

the needs of the whole product. Each team member is working on behalf of the whole team and the totality of the product. The Critique Meeting ensures the quality of the part in light of the whole.

Code review was one of the first structured review processes. A set of developers from the team walked the lines of code in a meeting to find bugs and structural problems early on.[6] Code review reduces the likelihood of post-release defects.[11] The code review process has evolved along with coding tools.[4] Now getting feedback on one's code is standard practice where once there was none.

Work product review is much broader than the code. To be responsive to the market, teams gather data from customers and get feedback on how well they are doing every time they ship. If teams are using user-centered design techniques and iteration with customers, designers get continuous feedback from users on product concepts, designs, prototypes, and minimum viable products. Teams also get feedback from internal stakeholders and other team members. Teams regularly review each other's work in code reviews, design critique sessions, storyboard reviews, and other internal review processes. Team members also get feedback from managers and senior professionals on their work individually or as a team. People on the team must deal with receiving critique on their collective and individual work over and over. Feedback and critique are a regular part of product design and development.

Much research on feedback and gender has focused on formal evaluations.[2] Unfortunately, developing good practices around the critique process has received little research attention.[8] A typical critique meeting involves a person presenting their work product to a group that criticizes it by identifying its known, or presumed, positive qualities and its known, or presumed, limitations—all of which may be debated. Following this feedback, the work is presumed to be improved.[3] These critique interactions may be in meetings or in one-on-one sessions. They may be more or less formal, having some or no well-defined process. They may be planned or come with no warning.

In the Introduction to this book, we shared that diverse teams that include women produce the most innovative and successful products. But also, that bias permeates organizational life, including critique practices. Women's voices may not be heard. Women's technical skill may be devalued. If women are alone on a team, less experienced, or in a job that does not have as much power

A good critique process ensures that all diverse voices are heard

as others, they may not be as confident or comfortable when speaking up. If we want product excellence, we must be sure to incorporate the voices of women and other diverse people in any feedback session. So how do we do that so that it works for all members of a diverse team?

The critique practice we share in this chapter has been used and iterated with real technology teams by both authors for over 30 years. We know that it works to get reliable feedback from all voices in the room that can then be incorporated into the product design. The practices and perspectives that guide the structure of The Critique Meeting provide a framework for how to run effective feedback sessions and how to interact when providing and receiving feedback on a work product.

Not all review sessions are performed in a group, but many are, however loosely organized. Organizations routinely run internal group reviews of storyboards, designs, product concepts, information architecture, interaction design, visual design, and slideshows. Sometimes outside reviewers, whether experts or customers, are invited. Having a clear process ensures teams get the feedback they need in an orderly way where all voices are heard, and no single voice dominates.

Our goal in presenting The Critique Meeting process[9] is to stimulate you to think about how you review your different work products.

BEYOND PERSONAL EVALUATION IN CRITIQUE MEETINGS

We all grow up with the idea of report cards and evaluation. Similarly, we are evaluated for salary and promotion through reviews. But if we want excellence in a work product, team members must separate work product review and internal critique meetings from the experience of being graded. A healthy feedback culture is focused on improving the product. The more we separate these two experiences the better. A well-defined critique practice goes a long way to ensuring that personal evaluation is separated from product improvement. A well-defined critique practice can ensure that professionals grow their skill with feedback and don't feel "graded." Feedback that looks like or can be perceived as a personal attack undermines these goals. And feedback that does not include all of the voices on the team, no matter their level of experience, power, gender, or other characteristics, simply undercuts the creativity and efficiency of the team.

So how we run our critique meetings and interactions matters.

ATTITUDES FOR AN EFFECTIVE CRITIQUE SESSION

A well-run critique meeting includes taking on a set of attitudes that help everyone focus on improving the work without undercutting the person. These include:

Improve the product—don't evaluate people. In a critique, an individual or a subteam presents the current state of the part of the work product they were tasked to do. A critique meeting brings the larger team together to provide feedback on the part and raise issues that may have been overlooked. When deep in the weeds of design, we may have overlooked a design standard, a customer need, or a need for consistency with a similar function in another part of the product. A critique meeting provides these wider perspectives so the sub-team can improve. Feedback is not a grade of the person; no one person or sub-team can think of everything.

Work is iterative—the presenter(s) is sharing a draft. The presenting team member(s) were asked to create an iteration of their piece *for the team*. The team is responsible for producing the overall work product. Implicitly, each sub-team is asked to produce an iteration—not a perfect final aspect of the work product. In a healthy critique process, reviewers recognize that they are reviewing an imperfect iteration.

[9] See https://www.witops.org/design-critiques/ for an introduction to the process.

They know that the presenters did significant work on behalf of the team as a whole, and they are grateful. Like with any early draft, in a group feedback session the sub-team is asking for critique from the larger team or outsiders to make their part better. To recognize the considerable effort needed to produce this iteration, we take time to explicitly thank the presenters for their work at the start of The Critique Meeting.

We are learning—not defending our position. To produce this iteration, the presenters made decisions based on their best information and skill. They have reasons for what they have done. If presenters come to a critique session looking to be told that everything they did is wonderful and that all the work is perfect, they will be disappointed. Instead, the presenters invite feedback, and they recognize they cannot improve their work without it. They are there to gain perspective and learn. A feedback session is explicitly looking for holes in the work product and to clarify anything that cannot be understood. Defending the current state of the work gets in the way of hearing the feedback, learning new principles or ways to design, and understanding issues the presenters need to consider. Presenters should not look at a review as a measure of personal worth or evidence for a performance review. The goal of The Critique Meeting is to collect new perspectives, and then use them to make choices to improve that part of the product. Indeed, everyone in the room will benefit from taking on an attitude of learning. Given the many perspectives in the room, all team members can gain from them, especially new hires and early career professionals. Taking on an attitude of learning is the easiest way to let go of any feelings of personal evaluation, which so often leads to defensiveness.

Inviting feedback encourages reviewers to find holes in the design

Valuable feedback is based on principle, not personal opinion. The goal of any critique process is to help the presenters, the owner(s) of the draft, create a better iteration. But they can only do that if the feedback is useful. It is hard to improve the work product if reviewers cannot say why something is a problem. Without, for example, customer data, best practice principles of coding or design, business goals, or other important criteria for evaluation, the recipients of the feedback have little idea of how to improve the work. A good critique process articulates criteria of goodness relevant to the work artifact. Without criteria of goodness or at least reasons for the comment, feedback can feel arbitrary. Arbitrary person-based feedback does not improve skill or the product. Opinion-based feedback implicitly asks the presenters to capitulate to the request or fight for their position. Neither of these responses will improve the product or the person's skill. Articulated criteria of goodness and an expectation that reviewers share why their comment matters are necessary to mitigate interpersonal friction. A well-run critique meeting asks all participants to reference principles and criteria which underlie their suggestions or criticisms. Even if someone shares their "gut feel" we ask them to explore why it is a problem; without a "why" the sub-team cannot weigh the importance of the point.

Feedback that includes reasons for an opinion can be heard and acted upon

Straightforward respectful feedback can be heard. Even with criteria of goodness, the person who did the work may react to feedback as a personal evaluation. They may fear for their jobs or promotion. They may feel that comments denigrate their skill. They may be listening to the words of reviewers to see if they are valued. How people receive feedback depends on their self-expectations and their experiences. But it also depends on how the feedback is delivered. The women we interviewed value straightforward feedback that helps them improve the work product. When feedback is delivered without overtones of evaluation, without blame, and without overt or implied devaluing of the person's skill, it is easier to take. A good critique process encourages impersonal, straightforward, data or principle-based comments. A good critique process manages the way feedback is delivered. In a good critique meeting, everyone is expected to behave respectfully and professionally. An effective critique process has *rules of engagement* defining appropriate and inappropriate behavior that can then be monitored.

Meetings and interactions including feedback work best when all participants have the right mindset to give and receive feedback. Starting the meeting by stating these expected attitudes reminds everyone to show up professionally. The Critique Meeting is not a personal evaluation, but we all can still hear it that way. To become a constructive feedback culture, we need to set the tone for that to happen.

PERSPECTIVES ON PROVIDING FEEDBACK ON WORK PRODUCTS

In addition to keeping in mind the attitudes described above, giving and taking feedback is best guided by some key perspectives. Any feedback on a person's work products is only effective if the person receiving feedback is open to it. If they are not receptive, the learning opportunity is wasted. So, it is best for people to ask for feedback on their work and state what they want feedback on. If managers or others provide unasked-for feedback, the recipient may feel personally evaluated. And timing matters, the person may be

Be sure a person wants feedback on their work— or make it an expected part of the team culture

preoccupied with something else they must complete. For these reasons, make feedback discussions a standard part of the team's processes and a standard part of one-on-one meetings with managers or buddies. Also, when group critique is built into a team's practice for skill growth, everyone participates—no one is singled out. But like all feedback, it must be constructive and delivered with an attitude of valuing and support. If not, no one will listen.

Also, giving and receiving feedback is especially important for growing early career professionals. New hires want and need frequent feedback on how they are doing and on their work. They want to know they are doing the work the right way. So, they might ask a lot of questions. Willingness to answer questions and provide frequent feedback by buddies, the manager, and senior professionals is important both to ensure skill development and to communicate value. New hires (and other team members) might not ask questions or actively seek feedback. Still, they need it. Giving feedback in these cases can seem personal, so framing the conversation as coaching a valued employee is important.

Last, all statements about the work of another person communicate a person's value. The attitude and tone of the reviewer implicitly send a message about the person's skill. When everyone on the team takes and gives feedback well, the team builds cohesion and belonging. When

Feedback with implicit negative messages about the presenter's skill undermines performance

frequent feedback done well becomes part of the team culture, no one is singled out as needing more feedback than anyone else. When feedback is given clearly with the intention of improving skill and giving positive encouragement, we communicate that we believe in the person's potential. Ensuring the person whose work is being critiqued feels valued and has a sense of belonging to the team is always an underlying goal of any feedback session.

The Critique Meeting process we share is grounded in these attitudes and perspectives. Armed with this point of view, we now need a structured practice, one that is clearly defined with explicit expectation. An explicit, well-structured critique practice helps teams get the best feedback from all voices—and can interrupt bias and negative interactions.

THE CRITIQUE MEETING PROCESS

So, what does a good critique meeting process look like? A group critique is a working meeting. As such its success, like all working meetings, depends on having clear goals, roles, processes, and expected outcomes. In other words, the meeting has structure so that everyone knows exactly what to do. Structure ensures that a working meeting achieves its goal efficiently without interpersonal chaos.

Structured processes in feedback sessions interrupt interpersonal chaos and save time

The authors know from our own experience, our research, and the literature[1] that structured meetings work. When teams call us because they have devolved into chaos, contentious conversations, and time overruns, we always find that they are not following the practices for that particular working meeting. Moreover, structure is essential for remote meetings, which do not have as many non-verbal cues. A key finding from our remote work research is that teams needed to increase the structure of their remote working meetings: agendas, time-boxed discussion, roles, tools to share artifacts, and strong participation processes to manage conversations. The Critique Meeting, being so central to making products and so fraught with feelings about feedback, benefits from clear, explicit, and articulated procedures.

The Critique Meeting process we share here is a straightforward practice with clear goals, roles, procedures, and documents or artifacts that represent the work product to be reviewed. It starts when a sub-team or person requests a review. Review may also be part of the overall team design and development practice. A person's work may be nominated for a recurring team critique meeting that is focused on sharing and learning from each other. Or review may be the standard practice for ensuring overall product and design excellence. In all cases, the person(s) who did the work wants to or agrees to be in a review process.

The Critique Meeting is run by the owner of the meeting. This may be the presenters (the sub-team) or the organizers of a standing group critique session. The owners are responsible for clarifying expectations and roles, inviting the right people, finding a time, setting up a physical or virtual space, and organizing access to all collaborative tools to be used. We recommend that The Critique Meeting occurs in a limited timeframe of 1–2 hours. The meeting works best with 1–2 presenters and 3–6 reviewers. We recommend that participation is facilitated to ensure that the meeting procedures are followed, good feedback is received, and the meeting fits within the designated timeframe.

The meeting owner starts the meeting by framing the review as improving the overall product, not evaluating the person or sub-team presenting. All are asked to take on an attitude of learning, give feedback respectfully, refer to criteria of goodness for any feedback, and to abide by the rules of engagement. Reviewers begin by expressing appreciation of the individual's or sub-team's initial work on behalf of the team. Implicit is a promise of support or help if needed. Stating the attitude appropriate for a critique helps get everyone in the right mindset for a productive meeting.

Run The Critique Meeting as described below. Running The Critique Meeting is not complex or hard to do. Once a team gets used to the practice, getting ready and running it will become easy, like all team practices used repeatedly. We discuss the key elements of the meeting structure next.

Preparation	Procedure to Run the Critique Meeting
Before the meeting: • Invite reviewers 3–6 is an optimal number unless it is a weekly team critique. • Share with reviewers the goal of the critique, the design artifact being reviewed, criteria of goodness, rules of engagement, and procedure to collect feedback. • Make sure all reviewers have the context they need to understand the product artifact being reviewed. • Assign roles to run the meeting and explain responsibilities. • Get a pack of sticky notes for all participants. Or use collaborative tools like a virtual whiteboard. • Print or prepare the collaborative tool to display the material listed below. **During the meeting display:** • The physical or digital design artifact to be reviewed. • The goal and type of feedback desired from reviewers. • Business goals relevant to the product. • Key user data related to the larger product. • Principles of excellence for the desired level of design to guide evaluation. • Rules of engagement.	1. Remind the team of the attitude to take during the critique. 2. Thank the presenter (design sub-team or person who made the artifact) for their work. 3. Presenter states the goal of the critique, reviews the procedure, and shares rules of engagement. 4. Presenter walks through the design artifact piece by piece, explaining each piece. 5. After each piece is presented, reviewers can ask clarifying question in round-robin style. 6. If the presenter or other sub-team members can't answer a clarifying question, it is recorded as an issue. 7. Following clarification the presenter walks through the pieces again, collecting feedback in round-robin style while sharing 1–2 points at a time. 8. Reviewers must state positives before negatives. 9. Notetaker captures positives and issues as they arise and posts them on that piece of the artifact. 10. If anyone has a design idea (DI) of how to address an issue they write it on a sticky note and post it on the applicable part of the artifact, without sharing it. 11. Moderator makes sure that the meeting is focused on feedback and solutions are not discussed. 12. After the first artifact is completed, collect any additional or overall feedback in a third round of feedback. 13. If there is time, move to a second artifact and repeat.

ELEMENTS OF THE CRITIQUE MEETING

Each element of The Critique Meeting and step in the process is designed to keep the meeting focused on gathering feedback productively. The design of any critique meeting must be useful to the sub-team or person who created the artifact so they can then use the feedback to improve their work product. We discuss each element in turn below.

GOAL OF THE CRITIQUE MEETING

Clarifying the goal of The Critique Meeting is critical to its success. The goal aligns all participants on the purpose of the meeting, defining what is on or off topic. Without a clear goal, the meeting can stray to side conversations and disagreements on process that elongate and defocus discussion. This frustrates all participants. So, every meeting needs a *mainline conversation*, a clear statement of what is on or off topic. The mainline conversation defines both what topics are included in the conversation and those that are excluded.

The overarching goal of any critique is for the presenters to receive feedback from others so their work product can be improved. Depending on what artifact is being reviewed, the presenters will define further goals describing the type of feedback they are looking for. Unless explicitly excluded, the goal always includes how the product fits within the larger product requirements: the overall structure of the code, corporate standards for interaction or visual design, cross-product consistency, business goals, modern design standards, etc. The presenter sets the scope of the review and the type of feedback to be shared.

The review may also, or primarily, be to increase the skill of the person presenting their work product. In this case, the meeting structure is the same, but reviewers are also coaches building up the presenter's skill. A weekly review meeting with people in the same job role helps both the presenters and the attendees align their work practices, clarifying design approaches, for example. Accepted practices of how to write code, design an interface, or even layout a slideshow will change depending on the organization. Having a weekly meeting to share work helps new hires and early career professionals learn what is appropriate for that organization. Even if the goal of the review is educational, The Critique Meeting process is the same.

In all cases, the mainline conversation of The Critique Meeting focuses on providing constructive feedback on the artifact being reviewed. That is what is on-topic. Of course, reviewers may have design ideas as feedback is given or which simply pop to mind. But resolving issues or choosing a new design direction is expressly not a goal—this is off-topic. The purpose of asking sub-groups to do part of the work is so the time of the whole group is not absorbed

Critique is to collect feedback—not to discuss resolution

by every decision. And in the limited time of a critique meeting, there is no time to both collect important issues and come to a joint solution. If the group stops to find a shared resolution for each issue, the purpose of a critique meeting will not be achieved. More importantly, a brainstorming, visioning, or resolution meeting has a different goal, roles, criteria of goodness, rules of engagement, and procedure.

But, since generating possible solutions is inevitable, and more importantly may contain good ideas for resolution, The Critique Meeting includes a procedure to stay on topic. People with design ideas are asked to write them down for later reference by the sub-team. Participants explicitly do not discuss the ideas. Writing down ideas is a good way to get ideas from everyone—and to avoid defocusing the meeting or allowing a vocal or influential person to dominate the meeting.

The mainline conversation of a critique is to collect feedback on the displayed artifact at the level of design desired to improve the overall work product and/or for training. Articulating the goals of the critique allows the participants to be sure they are contributing value. Articulating what is excluded from the critique helps everyone participate productively. Goals also allow the moderator to ensure that everyone stays on the mainline conversation. Without explicit goals, any meeting will wander and fail to achieve its purpose. And when meetings devolve into interpersonal chaos, they open the door to bias.

ROLES IN A CRITIQUE

All well-run meetings need clear roles along with associated behaviors. These roles are assigned before the critique and reiterated at the start of the meeting. Roles for a critique include:

Presenter. The presenter is one member of the sub-team who understands all the parts of the artifact, its overall design, customer data, business goals, and design principles behind the sub-team's choices. If the artifact has two clearly separate pieces, there can be two presenters. But to ensure smooth conversation we do not recommend multiple presenters for parts of the same artifact.[10]

Notetaker. The notetaker is another member of the sub-team. If the presenter worked on the artifact alone, they ask another team member or colleague to be the notetaker. The notetaker captures the feedback in the manner the team agreed. Use real or virtual sticky notes to capture each issue and place them on the part of the artifact the issue addresses. The notetaker is not taking minutes in a separate document; the feedback note is attached to the relevant place on the artifact. This makes it easy for the team to later deal with the issues during resolution. And because issues are posted, reviewers can see that their point was heard and recorded.

Reviewers. Reviewers are the invited people who will perform the review. A good number is 2–4 reviewers and never more than 6. The presenter and the notetaker are not reviewers. When reviewers come from the team or the team's management, they

---

10 A sub-team from a product team should be two people or at the most three. But in an educational setting, student teams are often 3–5 people. Each student is learning to present and how to take feedback. In this case, over the course of multiple reviews switch out the two presenters so everyone gets a chance. Or if you have time to review multiple discrete artifacts—then three presenters will work.

should understand the overarching goals of the project. Reviewers from outside the team need preparation before they can participate meaningfully. The sub-team can plan a pre-meeting, a 5-minute video, or some other means to introduce the project. Outside reviewers are useful for getting feedback from collaborating groups and in weekly feedback sessions to enhance the skill of the team.

Moderator. The moderator ensures that the meeting stays on the mainline conversation, the agreed-upon procedure is followed, all voices are heard, and the rules of engagement are adhered to. A good moderator has the gentle but authoritative skill to keep people on track and stop behavior that is not appropriate for this meeting. The moderator can be another member of the sub-team or another team member. The moderator is not a reviewer while they are moderating.

Before participants attend The Critique Meeting, they need to know the role they are to play and what is expected of them. Articulating roles helps everyone know what to do and gives the moderator permission to keep the meeting moving along smoothly.

FOCUS OF THE CRITIQUE: THE ARTIFACT BEING REVIEWED

The central focus of a critique is always the tangible artifact to be reviewed. This may be the code, user research representation (affinity diagram, journey map, work model, persona), user interface wireframe or other design artifact, storyboard), visualization of a concept, or slideshow—any work product. (See examples, Figure 8.1 and Figure 8.2.)

Before the meeting, everyone needs to know what is being reviewed. The artifact being reviewed must be tangible in some way so that it can be displayed and systematically walked through. Lucy

Choose the level of product design to review

Suchman[12] discovered that when product teams think together, they hover over some external representation—the code, design or research representation; data model; visualization of a concept; slideshow; or a napkin sketch. The artifact to be reviewed further defines the scope of the critique. The artifact also defines the scope of the feedback desired. For example reviewers examine different aspect of the work product in each of these different artifacts:

- **User data representation.** The clarity and level of detail of the representation appropriate to the type of data synthesis; the visual representation showing key distinctions; overall usefulness.

- **Vision** (concept representation). The high-level concept and how well it meets the needs of the users and the business; what is technically implementable; level of detail—too much or little to communicate the concept.

- **Storyboard.** The quality of the practice proposed; the set of steps that we expect people to do to achieve their intent in the new design or product.

- **System structure or information architecture.** The proposed structure, function, and flow of information and action within the design; each place in the design supports a clear intent.

- **Interaction design in wireframes.** The structure, flow, and consistency of the function and information in the user interface on and between screens.

- **Visual design.** The graphical presentation of the interaction design and how it represents the brand.

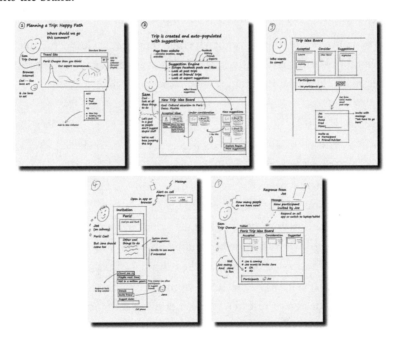

Figure 8.1: An example storyboard.[11]

11 This storyboard example was published in: Contextual Design Second Edition, Karen Holtzblatt and Hugh Beyer, pages 316-18, Figure 13.1, Copyright Elsevier 2017.

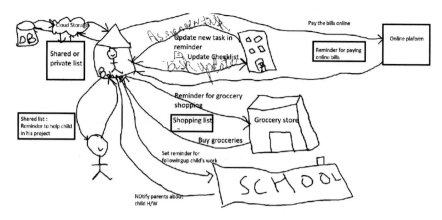

Figure 8.2: An example vision drawing.[12]

Each product artifact defines a different level of design which further focuses the review. Each artifact also implies what not to review. So, for example, if we do rough user interface drawings in a storyboard, we must explicitly state that we are not looking for feedback on the UI layout itself. If we display a prototype or piece of running code, we must be very explicit about what to review. Any running piece of code represents all the levels of design as well as the quality of the code. In this case, the sub-team specifies what they want reviewed or they will get unfocused feedback. The artifact and clarification of what the reviewers should focus on ensures reviewers know what to pay attention to in The Critique Meeting.

Display the artifact so everyone is focused on the same thing during discussion

The artifact also gives the meeting a visual focal point, a map to traverse in the meeting. All reviewers will be looking at and responding to this one artifact. The meeting works best when the artifact is displayed for all to see and reference at the same time. Hang it on a physical wall in the meeting room or share it in a remote meeting. Then be sure everyone can see the pieces of the artifact that are being discussed. Giving everyone access to their own copy of the artifact can be useful—but not for running the meeting. A productive meeting requires that everyone is focused on the same thing at the same time.

CRITERIA OF GOODNESS

Criteria of Goodness define the principles of excellence and issues that are important for product success. We recommend displaying three separate lists of criteria to keep the review team focused on all three areas: principles of good design for the artifact, business goals, and customer practices, issues, and needs. These areas represent key dimensions important to consider in any review.

[12] This visualization of a product possibility comes from Nicola's students: Anusha Spandana, Burak Domac, Parth Shah, and Rasheed Jadallah.

> ## Principles for Goodness: Interaction Design
>
> **Primacy:** Everything needed and only what's needed is represented in the largest, central place.
>
> **Immediacy:** The system "thinks" for the user, providing options or information at the top. The user doesn't need to go looking for things.
>
> **Visual Flow:** The user can scan the screen and information and immediately knows what is being communicated and what to do.
>
> **Complexity:** The screen presents only information and function needed for the intent—nothing more.
>
> **Consistency:** The UI mechanisms and page layouts are consistent across the system.
>
> **Learning:** The system builds upon the user's knowledge of standard ways that technologies, UIs, and products interact and operate.
>
> **Modernity:** The design uses interaction patterns considered "modern" or standard.

The criteria of goodness for any artifact depends on the type of artifact being reviewed. Principles of good design for code are different from those for affinity diagrams, journey maps, visions, storyboards, interaction, or visual designs. Principles of goodness can also include usability and accessibility standards. For example, see the criteria of goodness the authors use for UI interaction design.

Criteria of goodness also include business goals. Too often the business goals for a product are not articulated. But when business goals are not supported, products may be canceled, or the team may get harsh feedback because their work does not support the business needs. Any work product needs to be aligned with the business mission, timeframe for delivery, the team's ability to deliver, and the financial business case. Listing business goals explicitly helps the team and reviewers keep business needs front and center.

Last, for product success, the product needs to align with and support what works for the customer. Hopefully, the company is using a customer-centered design process and has collected data on users' practices and potential delighters. Using that data, the top user issues can be prominently listed during the review to keep the user front and center for reviewers. Feedback can then include whether or not the work product supports or undermines a user issue or practice.

The exact criteria of goodness depend on the artifact, the company, and the team. Articulating criteria of goodness reminds the team that feedback must include why it matters and what the impact is for the business or the customer. Feedback may also point out violations of current best practices in design. All comments on what works or doesn't work about the artifact can then reference the associated criteria for goodness. If the reviewer has an opinion that is not related to these criteria, encourage them to reflect upon and identify an underlying unarticulated criterion that can be added to the list. If no "why" can be found for an opinion, mark it as opinion. Criteria of goodness help the team move away from sharing their unsupported opinion, which can so often pit one person's idea against another. The presenters need to know the basis of any critique to guide their decisions when creating the next iteration.

PROCEDURE TO COLLECT FEEDBACK

Feedback is collected during The Critique Meeting in two rounds with clearly stated rules of participation. If the meeting has no clear structure around how people share their ideas, participants tend to talk over each other, prolific or influential participants can dominate the sharing, quiet people may not share, and women and other underrepresented people are less likely to be heard. This is also true for new hires and early career participants. As we have said, how the team collects feedback matters so that all diverse perspectives come to bear on the quality of the work product.

Round-robin participation ensures that all people and perspectives are heard

The Critique Meeting comments are collected in round-robin fashion, moving from reviewer to reviewer in order. Each reviewer shares 1–2 items when it is their turn, allowing others to share their ideas before the first person shares additional items in a second go-around. Participants often find they are not the only ones with the same idea; that others have valuable ideas they did not think of. Collecting comments in an organized way on a work product also allows the notetaker to keep up.

If there is more than one artifact to be reviewed in the meeting, they are dealt with one at a time. Return to Round One and Round Two for the second artifact.

Round One: Clarification Questions

The presenter walks through the artifact describing one part at a time: a storyboard cell, a screen, a part of a journey map, etc. Reviewers listen and ask clarifying questions to understand the design and the presenter's intent. Clarifying questions are not feedback; they help everyone understand the design and structure of the artifact. Clarification helps everyone get on the same page. Often work products, particularly in early iterations, are incomplete or their representations are incomplete. For example, in a storyboard review, all steps may not be represented even when the sub-team has discussed them. The correction may simply be adding more cells for clarification.

Clarifying questions do not include hidden feedback. "Have you thought of using this interaction design approach?" is a design idea hidden as a question. Clarifying questions are for the purpose of understanding the design of the artifact and what the presenters were trying to do. Dealing with clarifying questions first ensures that reviewers get the context they need to understand the work artifact before thinking about feedback.

In Round One the presenter describes each part of the artifact, dealing with clarifying questions on each part. The presenter then moves to the next part and reviewers again ask clarification questions until all parts are shared. No feedback on the quality of the work product is collected until Round Two as described below. If the presenter, or other sub-team members, can't answer a clarifying question it is recorded as an issue. It is not discussed. Sometimes clarifying questions reveal "holes" in the work artifact, something left out or not thought of. The notetaker records the missing item as an issue for the sub-team to consider later.

Round Two: Feedback Referencing Criteria of Goodness

Once clarifying questions on the presented part are answered, reviewers may share feedback. Feedback speaks to what works and doesn't work about the work product based on the criteria of goodness and review scope. The presenter returns to each part of the work product, soliciting feedback on each part in round-robin fashion. Or if the work product is small or very coherent, reviewers may share feedback about the work product as a whole. The moderator can help participants determine the best approach.

Two classes of comments are welcome in The Critique Meeting. All other comments or discussions are considered off-topic. The two classes of allowed comments are:

- **Feedback: validations, issues, and holes.** Reviewers are asked to look for positives in the work product, how it supports the criteria of goodness, as well as problems. To encourage this, reviewers are asked to share positives before they may give negatives. After sharing how the work product works well, reviewers may share issues that violate criteria of goodness or holes the team needs to consider. The notetaker captures each issue on a sticky note and posts it on the relevant part of the artifact.

- **Design ideas.** Any participants may hear an issue and generate a design idea of how to address it. Everyone in the meeting is given their own sticky notes (physical or virtual). If they have a design idea, they write it down and label it with DI. They post the DI on the relevant part of the artifact after the feedback round for that piece. Later, the sub-team can consider these ideas in their next iteration. Design ideas are not shared with the whole group in order to discourage the group from segueing into a redesign meeting. Resolution of any issues is not the focus of The Critique Meeting.

Feedback may offer an alternative as a means of expressing how something might look if it adhered to a principle of modern design, for example. In this case, the issue to be recorded is the violation of that principle of modern design; the idea offered for how to correct or improve it is a design idea to be recorded. The goal of managing participation in this way is to keep everyone focused on providing the presenters with the feedback they need, but then trusting them with resolutions. The sub-team knows that later they can come to members of the team for help in the redesign if needed. So, in the meeting the moderator helps participants clarify the issue succinctly, reference criteria of goodness, and record it.

We encourage reviewers to write their feedback issues in Round One while listening to the presentation so they are ready to provide feedback in Round Two. This allows each person to clarify their own thinking without being influenced by others. It also invites quiet or less experienced people to develop a point of view. Review comments are then shared in round-robin fashion during Round Two. Each reviewer first shares 1–2 positives and then may share 1–2 negative issues. The next reviewer follows in the same fashion. In the second go-around, each reviewer again shares 1–2 issues or passes if they have no more comments that have not been covered by others. Two rounds should be enough, three is certainly sufficient. Round-robin sharing can also help modulate participation so that no one person dominates the meeting.

Moving from reviewer to reviewer quickly allows all to speak. The other reviewers with the same issue know that it has been recorded, they do not need to share it again. Moving from reviewer to reviewer ensures that no one is waiting for their turn or to see what influential people say before expressing their own thoughts. Stating an issue and writing it down without looking for others to comment or agree also ensures that each reviewer is heard without criticism. These practices encourage all reviewers to form their own idea of ways to improve the work.

Online collaborative tools may change the process of a review meeting. When online, it is easier for people can enter their feedback and design ideas simultaneously. Some think this saves time and allows quiet people to voice an opinion without speaking in front of a group. Some like to enter comments before The Critique Meeting. But whether or not people review the artifact before the meeting, we have found that feedback is more valuable and less repetitive if it is shared in-person as we have described in Round One and Round Two. Then each reviewer can get and hear clarifications before forming an opinion. Issues raised are then more relevant than they are when people form opinions about the artifact with no clarifications. In Round Two, reviewers can hear each other's responses without discussion, which increases a shared understanding and avoids duplication of issues.

Sharing feedback verbally helps clarify issues and context—and reduces redundancy

So even if you choose to run your meeting with simultaneous input by reviewers, entered the day before or in real-time, we encourage that everyone shares their clarification questions and feedback to the larger group in Round One. Then reviewers who had entered feedback before the meeting or while hearing the presentation can refine their notes. In Round Two, as everyone listens to each other, participants can skip (and delete) their own comments if someone identifies the same thing first.

In a Critique Meeting, the process for collecting comments from reviewers matters to ensure that all understand the work product, that all voices are heard and considered equally, and that off-topic discussion does not derail the meeting. The Critique Meeting is a core process that communicates the value of the sub-teams' work and skill. Managing the tone and way feedback is collected is central to ensuring that women, new hires, and all voices are heard. These techniques for managing feedback also work in other meeting types—decision-making and brainstorming meetings, for example. But these practices for sharing only work when the overall process is structured and guided by rules of engagement.

The Critique Meeting Rules of Engagement

Attitude
- Improve the design, don't evaluate people.
- Help everyone learn.
- Use professional behavior.

Thank the design sub-team
- Acknowledge their work on behalf of the larger team and overall product.
- Assume they did their best.
- Remember that this is one of many versions.

Start with positive about what works
- If you have negative feedback, first state positive feedback.
- Don't throw the baby out with the bathwater.

Reference the reason for your response
- Use user data, best-practice principles, technical feasibility, brand support, etc. for any feedback, positive or negative.
- If you are responding from your personal experience, link that opinion to existing criteria of goodness or identify a new one.
- If you cannot reference a reason, mark the opinion as personal.

No discussion
- Share positives, feedback, or issues only.
- We are not here to resolve issues or redesign the artifact.

No defensiveness
- We believe in the value of all participants; you do not have to defend your choices.

Listen to the moderator
- We agree to allow the moderator to modulate our participation according to these rules of engagement and the meeting procedure.

RULES OF ENGAGEMENT

Each company may have its own overarching rules of engagement. When done well they encompass expected ways of behaving that ensure respect for all. Agile teams may also have their Team Manifesto declaring how they want to interact. (See Chapter 11 for an adaptation of this for all types of teams.) These are all good practices and people should be reminded of them often. But specific types of meetings also benefit from their own particular rules of engagement. This helps everyone know what is allowed and what is discouraged in that type of working meeting.

See the rules of engagement we use for The Critique Meeting. Feel free to add your own as appropriate to your context. Display the rules of engagement on the wall along with the criteria of goodness and the artifact. In a remote meeting, display them on a virtual wall. Also, remind participants of the rules of engagement at the beginning of the meeting; read them aloud. Sometimes teams ask participants to sign or commit to the rules formally.

Having explicit rules of engagement for different kinds of meetings raises expected behavior up to consciousness. This also gives moderators the power to enforce this expected behavior. Without explicit acknowledgment, moderators may feel hesitant to manage co-workers' behavior, especially that of more influential people. The moderator can then reference the rules of engagement when gently but authoritatively correcting behavior.

THE POWER OF STRUCTURED MEETINGS

Critique is at the center of every maker industry. If we want to become a productive feedback culture, we must ensure that people can hear the feedback so they can improve their skill and their work products. To do this, any critique process must ensure people treat each other professionally. Sharing feedback by referring to criteria of goodness—not a person's skill or failing—helps set the tone needed for a professional valuing culture. All voices are heard and considered; we build product quality and team cohesion.

But if the meetings are not well-managed, chaos and inadvertent devaluing of others can happen. When a meeting does not manage participation, people are more likely to talk over each other, one speaker can take over the meeting, and the meeting goals may not be achieved. When power is not leveled ensuring that everyone's perspective is equally valued; when distain tinges feedback, we undermine team cohesion and innovation. Or, when we devolve into debating the best resolution of an issue instead of providing valuable feedback, we waste everybody's time. Without well-run working meetings, we risk alienating women, new hires, other diverse people, and anyone who is dedicated to reaching good results. We undermine the promise of diverse teams.

Structured meetings designed to include the practices of well-run meetings counteract this all. The Critique Meeting process presented here represents an example of what is needed for any well-run working meeting:

- Meeting owners who communicate clear goals and expectations for the meeting

- Well-defined processes including clear goals, roles, procedures, and expected outcomes

- An artifact as a focal point for all to "see" what is being created or agreed to

- A small enough number of people so everyone has airtime and/or ways of managing participation

- A moderator for groups of > 3 people

- Articulated rules of engagement relevant to that working meeting

- Shared criteria of goodness relevant to decisions being made

Too often in organizations we resist structure. We think that structure places a burden on creativity. But this leaves any working meeting open to interpersonal chaos. Without structure, every team is at the mercy of the natural skill of whoever happens to be on the team. Do you have someone who continuously moderates the conversation to keep it on track? Do you depend on that one person who knows how to refocus the meeting when the voices get louder and conversation becomes less productive? Do you happen to have that person who is sensitive to how everyone is reacting and steps in to smooth over conflict and potentially demeaning comments? We can hope that we have put together a perfect mix of people. Or we can design our working meetings, making them more likely to succeed no matter who is in the meeting ensuring that diverse teams really work for everyone.

The elements and the procedure of The Critique Meeting shared here help everyone know what is expected and what to do. When a structured critique meeting becomes standard practice, preparation and smooth participation are not hard. Everyone will know what to do because it has become the standard team practice. With clear structure, team members receive feedback that considers all perspectives, can be heard without feeling judged, and includes principles that can guide the redesign of the work product.

More importantly, structure in working meetings teaches everyone how to give feedback. The practice of professional giving and taking feedback gravitates from formal critique meetings into interactions with managers and co-workers. Of course, this does not guarantee that people will always be respectful, patient, and professional in a structured meeting. People will always have the ability to derail anything. But with clear expectations and processes, it is easier to stay professional. And professionalism is at the root of what all diverse people and diverse teams need in everyday work—indeed what all workers need to really get things done.

Structured critique meetings can teach best practices for giving feedback in groups or one-on-one

The principles and processes we share here are about how to run The Critique Meeting, but they may be adapted for any working meeting. For example, the working meetings in Contextual Design, a user-centered requirements and design process, are also structured based on these principles.[9, 10] Find and design working meeting processes that work for you and your teams.

With explicit structure, everyone thrives. So, it is not surprising that the industry has increasingly moved toward Agile processes—the most highly structured software development process used in the last 30 years. In the next chapter, we explore Agile and how to tune it to work better for women and diverse teams.

CHAPTER 9

Sneak Attacks on Key Processes: Agile

At the core of our redesign approach is the principle of sneak attacks. We look for small changes to existing practices central to both improving the work of diverse teams and the experience of women. The Team Onboarding Checklist is just a checklist. Yet, it is a tool to help clarify and improve an otherwise undefined onboarding process that also improves the work experience for women. The Critique Meeting procedure adds structure to how people share feedback ensuring that all diverse voices are heard, and that the team gets high-quality feedback. The Quick Reference Guide is a card that managers can put on their desks—a little reminder of best practices in onboarding. These recommendations are simple process improvements. They represent sneak attacks on process change, with an eye toward gender issues. In this chapter we use Scrum to further explore how to create sneak attacks on key processes.

Small changes to key processes can interrupt bias and improve performance

Bias is hard to eliminate—but it is not hard to interrupt.[30] Changing onboarding and critique simultaneously disrupt stereotypic attitudes and behaviors, support what women need, and improve the process for everyone. They do not announce that they target gender issues. They sneak in important changes that improve the work environment for diverse teams and women. To redesign practices to help women thrive, slide tiny changes into existing practices so that they aren't very noticeable.

But what sneak attacks on process should we take up? Central to the success and impact of our intervention recommendations so far is to make an implicit or undefined way of working explicit and chosen. Implicit practices open the door to bias and stereotyping. What we do out of habit and without reflection cannot be examined either for effectiveness or for inadvertently supporting biased behavior. Explicit practices and expectations can be designed, examined, and monitored. Sneak attacks on implicit practices are one of the most effective interventions to retain women in tech. To start, we must find the implicit practices and expectations in a key practice.

To help managers and teams explore implicit and explicit elements of key practices we use The Analysis Matrix. We use Agile, the most widely used software development practice today, as an example of how to find and create sneak attacks on a process. We share small practice changes that you can sneak into your processes to interrupt bias. Let's start by understanding Scrum, the most popular form of Agile.[17]

KEY DIMENSIONS OF SCRUM

I like Scrum. We now have a daily stand-up where we go around and everybody says what they worked on the day before, what they are working on today, and whether there are any impediments. It's becoming obvious who hasn't been working—it's starting to be embarrassing for some people. Also, with the empirical approach and ongoing feedback, negative behaviors are being identified and changed—it is actually very good. Michelle, Solution Architect

Michelle likes the changes introduced by Scrum. Its practices, she says, highlight what is working and what is not in the way they work and collaborate. In other words, many of its practices are explicitly defined and guided by principles. This helps the team make improvements. Clarity and continuous improvement of the practice are indeed some of the guiding goals of Agile. There is much that works about Agile, but can we tune it to make it even better?

The "Manifesto for Agile Software Development"[6] was put forward in 2001 by 17 male software developers who had become disenchanted with the current practices. They eschewed the waterfall development technique, and a myriad of other approaches, which sought to get the requirements and design right before building. Agile's founders worked on business and government systems where functions had to satisfy stakeholders in the business. Agile has now been adapted for commercial products, website development, and beyond. To help our discussion we summarize some of the key practices of Scrum below.

Scrum developers deliver an iteration, a chunk of functionality representing a sub-set of the product, within a fixed timeframe of 2–4 weeks that are called sprints.[19] The Scrum team[13] organizes what they need to do in a series of well-defined working meetings driving them to deliver the iteration within the sprint timeframe. Companies like it because short sprints are more likely to deliver something within a specified period of time.[29] Not all companies use a 2-week sprint, often it is 3–4 weeks. And not all companies ascribe to all of the tenets of Scrum, but they do use most of them.

Compared to other software development processes, Scrum is highly structured. Key team roles are defined (See Scrum roles). Scrum also defines how each working meeting is run, specific artifacts to be used in each meeting, and processes to guide how developers code (See Scrum working meetings). Individuals are left to write their code, but their piece must fulfill its purpose and meet coding standards. The team's Definition of Done, a shared understanding of how the product increment must behave to be released, guides release. Developers must resolve any comments from

---

[13]   Agile's concept of team focuses on the developers doing the coding. Notably absent are user researchers, interaction designers, content specialists, and other cross-functional roles. Fitting them into the team process is the subject of much discussion. In addition, the Customer role does not include users of the potential product and makes no distinction between users and stakeholders. We are not addressing the process limitations of Agile techniques from a user-centered design perspective.

code review and be sure the code passes all tests. Also, Agile calls for home/work balance—work at a sustainable pace—no crazy all-nighters for weeks on end.

Scrum Roles

1. **The Team.** The set of developers responsible assigning work tasks and doing the work.
2. **Product Owner.** Defines the properties of the product to be developed in a sprint. Often operates like a product manager modulating between the needs of the business, user requirements, and technical challenges in a sprint. In practice, keeps people on track.
3. **Scrum Master.** Moderates, teaches, and coaches on all agile techniques. Acts as a facilitator. Although often outside the team, may also be a developer on the team.
4. **Customer.** This is a business stakeholder to whom the product is delivered. The purpose of the role is to give them a voice in an iterative process as they react to the "goodness" of iterations. This role substitutes for extensive requirements documents. The customer here is not the user.

Scrum Work Meetings

Agile is a series of working meetings. Each is structured by describing exactly what to do or say in the meeting. Often an artifact is produced or is the focal point of the discussion in the meeting.

1. **Daily Stand-Up.** Every day the team sets aside 15–20 minutes where each member answers three questions in turn: What did I do yesterday? What am I planning to do today? What blockers or impediments do I have to getting the work done? The stand-up may uncover a blocker that is then resolved by a sub-team in a subsequent meeting.
2. **Iteration Planning.** At the beginning of each sprint the team and Product Owner meet to determine what stories to include in the next iteration. Stories are small function scenarios displayed on story cards. All story cards are displayed on a task board. Participants write the goal of the iteration and then select stories they mutually agree on.
3. **Planning Poker.** The team must determine the cost in developer time of proposed stories to build in an iteration. Planning Poker is a technique to estimate the cost. The result may change what is in the iteration.
4. **Review Meeting.** At the end of the sprint the team reports on what they have done in the release. The review meeting is effectively a deliverable meeting where the team shows what they did and how it fares against the original goals, cost estimates, and acceptance criteria.
5. **Retrospective.** The Retrospective is at the end of the sprint. It focuses on evaluating the process used in the last sprint with an eye toward improving the process.

Scrum strips managers of their traditional responsibility for work assignment and oversight, giving those choices to the team. Responding to their past experience with managers, Scrum instead institutes principles that keep management from derailing the work. For example, all requirements and any changes are frozen during a sprint. During a sprint, the team focuses on building the agreed-upon iteration. Change can happen between sprints during the planning meeting. Similarly, no new development can be started until the current iteration passes all tests. This establishes the primacy of testing in the decision of what can be shipped vs. management's arbitrary desires.

Within this overall structure, Scrum teams self-organize particularly with respect to selecting what to work on and planning what should be in a sprint. The primacy of the team in managing the work is a central tenet of Scrum.

Our purpose here is not to introduce you to the Scrum flavor of Agile. Many books and consultants are available for that.[14] Nor is our purpose to talk about how to make Scrum a more effective practice; others have addressed the issues of the absence of practices to collaborate with user experience professionals and others who must work together.[7] Still others point out the negative consequences of having no overarching product concept, workable UI structure, infrastructure development, and documentation of any kind. Instead, we examine Scrum for ways to tune its practices to improve the daily work experience for women and diverse teams.

Research on Scrum and gender is sparse. Anecdotally some say that Scrum might be better for women,[14] mapping the Scrum principles to female work characteristics.[23] Some say Scrum is better for women because it leans on conversation to make decisions.[12] But then again, women do not always feel heard in conversation. Perhaps selecting your own tasks vs. assignment by managers overcomes the bias that women are not as competent as men. Planning Poker may help women's voices be heard because estimates are done individually before being shared. One interesting research finding is that the company's attitude toward gender diversity and related policy is more important than the actual number of women on an Scrum team.[3] Previous research and Scrum practitioners emphasize the importance of a good match between an organization's culture and Scrum methods for adoption to be effective.

We agree. Diversity-related change in Scrum techniques works best when it occurs within a larger corporate commitment to women and diverse teams. But overall, we do not have enough research exploring the impact of Scrum on women's experiences at work.

Scrum already uses practices that work well for diverse teams—but it could be even better

When we examine Scrum practices, they resonate with the kind of practices that work well for women, early career professionals, and diverse teams. From that point of view, Scrum is kind of amazing. It is loaded with well-defined goals, roles, processes, practices, time-limited interactions, and ways of getting the chaos of shifting requirements out of the overall process. Data and feedback

---

14 The Scrum Guide by Ken Schwaber and Jeff Sutherland,[27] co-authors of the Agile Manifesto, introduces the basics of Scrum. There are many helpful guides (e.g., Ilya Bibik's *How to Kill the Scrum Monster—Quick Start to Agile Scrum Methodology and the Scrum Master Role*[8]). Bertrand Meyer's *Agile! The Good, the Hype and the Ugly*[19] offers brilliant (and entertaining) analyses of the different Agile approaches. Jutta Eckstein offers valuable insights into linking Agile teams and organizational development in *Retrospectives for Organizational Change: An Agile Approach*[10] and *Company-Wide Agility with Beyond Budgeting, Open Space & Sociocracy.*[11]

are used for improvements. Central artifacts like story cards, the planning board. and burndown charts are used as focal points for conversations, as we recommend for the critique process. Scrum calls this visualization, using a physical manifestation of what is being discussed. Also, the Scrum Master is a process coach and moderates meetings to ensure they are productive. Everyone knows that success is having your code pass the tests. Scrum is a highly structured articulated process. It is not surprising that Michelle likes Scrum.

As we analyze Scrum using the Analysis Matrix, our recommendations for tuning are informed by our experience and our interviews with women on Scrum teams. This includes longitudinal research[15] we have done to improve gender issues within different companies using Scrum techniques. To get started we share how making implicit practices explicit can improve women's experiences on diverse teams.

EXPLICIT VS. IMPLICIT PROCESSES: DECIDING WHERE TO TUNE

A good way to start thinking about improving a process is to identify what is explicitly defined and what remains implicit.[25] Transparency, another important pillar of Scrum, means that how work is done has to be concrete and visible so that the team has a common understanding. So, Scrum uses explicit day-to-day activities and artifacts to ensure transparency. Literature in gender studies, software development, social psychology, and organizational culture also analyze situations and surface potential issues by differentiating what is implicit vs. what is explicit.

When a practice does not have enough explicit structure and clear expectations it is prone to interpersonal chaos, as we saw in a typical critique meeting. Chaos unlocks the door to bias and unprofessional behavior; structure disrupts it. Well-designed, explicit practices mitigate interpersonal conflict, stereotyping, and bias. Moreover, if important processes are not well defined, they are unlikely to be consistently effective. If expectations

Well-designed explicit practices reduce conflict and stereotyping improving team effectiveness

and practices are not explicit, they are nearly impossible to examine or improve. And explicit practices ensure teams get the work done successfully. So, a good way to find what to target for improvement in a process is to look for places where too many expectations and procedures are left implicit.

For example, we were working with a company that had just started to use Scrum. Ranya used to be the Team Lead of a software team developing insurance software. With the adoption of

15   The research used a living lab approach[1, 2] to understand and create interventions to improve women's experience in technology companies. A synthesis of interview transcripts, field notes, and notes from informal interactions informed our analysis of Agile techniques with people in a variety of roles and ages. The research took place over a three-year period working with technology companies of divergent types. It was partially funded by the German Federal Ministry of Education and Research (BMBF), grant number 01FP1603, 01FP1604, and 01FP1605. The responsibility for all content supplied lies with the authors.

Scrum, her role was Product Owner. All of the developers and the Scrum Master were men. With that role change, Ranya's experience also changed.

> *Being part of the Scrum team gave me more insight into how the code was actually written. I noticed some terrible spaghetti code. I told the development team to clean up their act so that somebody else could understand, maintain, and expand their code. They kind of told me off, saying that it was not my job as a Product Owner to focus on architecture and code quality. Before, as a Team Lead, I could ensure that things are done a certain way. With Scrum and being the Product Owner, the developers think they are empowered to do what they want.*
> *Ranya, Product Owner*

From Ranya's perspective, after the role change the developers are behaving badly and the quality of the code will suffer. To help Ranya, we mapped out explicit and implicit aspects of this situation. Explicit aspects of the process include the Product Owner's defined responsibilities and the code that can be reviewed. Implicit aspects of the process include: the standards by which code quality is measured; who gets to address low code quality; differences in the perception of a Product Owner; and the myriad of personal attitudes, assumptions, and motives of individuals. As in all real-world situations, we cannot know whether the developer's aversion to Ranya's observation emerged from a bias against a woman telling the developers what to do, an unwillingness to put more effort in to rewriting the code so it could be more easily maintained and expanded, or something else.

Clarify desired behavior—do not focus on changing individual's implicit assumptions

Identifying the implicit and explicit aspects of Ranya's situation allowed us to define reasonable interventions. Together with the team, we started by making coding standards explicit and integrating code quality into the Definition of Done. We also defined and wrote down the exact role of the Product Owner in her team. We did not focus on implicit biases that developers or Ranya might have. Instead, we chose to interrupt any bias and improve the overall process with explicit standards and role definitions. Now the members of the development team understood that their code was going to be quality-checked. And Ranya understood that judging the product beyond what was visible to the end user was not part of her responsibility as a Product Owner. We did not have to bring up gender to improve how the team worked. We created a sneak attack by tuning the existing practice.

THE ANALYSIS MATRIX:
SURFACING GENDER ISSUES IN SCRUM

Our approach to helping Ranya's team shows how identifying what is explicit and implicit in a practice can improve both the effectiveness of the team and the experience of all diverse team members. The Analysis Matrix is a tool to help teams and organizations analyze their own practices.

In the matrix we separate individual and team practices and expectations to encourage reflection on the differing influences on each. Social scientific literature tells us that individual behavior is influenced by many factors that lie within the person, e.g., goals, motivations, attitudes, experience, and skill. But social dynamics also influence each person within a group or team context. Groups, for example, can create expectations and pressure on the individual to conform. The resulting four quadrants in The Analysis Matrix focus on helping teams identify implicit and explicit practices for both teams and individuals as targets for reflection and tuning.

QUESTIONS FOR REFLECTION

EXPLICIT

Team Practices

1. Are our working meetings and processes well-structured and articulated?
2. What procedures are still implicit, open to the influence of bias, power, and stereotypic attitudes?
3. Where can we introduce more structure to ensure both the success of the team and equal participation for all?
4. Where might our practices invite exclusion and devaluing of underrepresented and less experienced people?

Individual Behavior

1. Are the work expectations for each job role well defined and people coached?
2. Are criteria for goodness well-defined for each work product?
3. Are expected behaviors within working meetings clear?
4. Are challenging work and housework tasks pulled equally by all?

GROUP ←————————————————————→ **INDIVIDUAL**

Values and Culture

1. What are the values in practice within the group that regulate working meetings and daily interactions?
2. What is the culture of the team (e.g. habits, shared values, taste)? Does it work for women?
3. How does the team iterate how they work together?
4. How is the "perfect team" or "the good team member" portrayed? How gendered is this?

Beliefs and Biases

1. Have we designed daily practices with an eye towards undermining bias that help teams and managers collaborate and communicate value to all?
2. Do we have different expectations for people with different identity characteristics like gender? Who gets recognition and for what? Is it fair?
3. Who feels a sense of belonging and who does not?
4. Do we monitor how we work and treat each other to continuously improve the work life? Can women easily share their concerns and have them addressed?

IMPLICIT

To aid reflection on Scrum from a gender perspective, we created questions for each quadrant. These questions emerged from gender issues in literature, interviews, and observations of women working in Scrum encourage reflection. (See Analysis Matrix with questions for reflection.)

To use The Analysis Matrix to explore any practice, start by identifying explicit elements, placing them above the line. Then identify implicit elements and place them below the line. Do this in all four quadrants. To understand the power of the matrix and the opportunities to improve Scrum, we share our analysis and thoughts about Scrum. (See Analysis Matrix with results from reflecting on Scrum.) Try it out in our workshop with your group or team. Download a PDF guide here.[16]

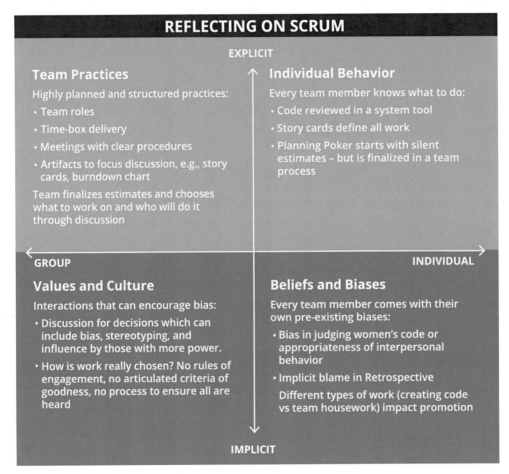

Scrum is already highly structured and much is very explicit. But could we do more? What is still implicit and therefore prone to chaos and decisions based on bias? In our analysis we identify

---

[16]   Download The Scrum Analysis Matrix workshop https://www.witops.org/scrum-matrix/. Contact us if you want access to our Miro board so you can copy and edit the workshop elements.

some of the aspects of Scrum that are explicit and those that are still implicit for both the team and the individual. This analysis is not meant to be an exhaustive list. Indeed, each key Scrum process and meeting type can be analyzed again by itself to uncover more see more potential for tuning.

Looking at our analysis we noticed that Scrum as a process targets team-based practices. The team holds meetings, makes decisions, abides by rules and roles, etc. So, if we want to tune Scrum with an eye toward gender, implicit practices within these team practices are the best place to start. Our analysis also shows that explicit individual behavior emerges from what people are supposed to do within working meetings and when coding. Both are highly defined and explicit. But individuals' implicit values, expectations, and assumptions are where bias is likely to be introduced. And we know that these implicit biases are most likely to emerge when group activity is unstructured or not explicated such as in decision-making discussions and work assignment. Using the Analysis Matrix helps us identify our tuning challenges: Redesign unstructured interactions to interrupt bias without annoying people with too much structure.

Central to Scrum values is *self-management*. The team is the primary decision-making body assuming that through conversations all issues can be worked out. But as we have said in Chapter 8 on The Critique, group discussion and decision-making in any working meeting are by their nature implicit and freewheeling, a context ripe for interpersonal conflict and bias. Scrum emphasizes the need for a *Scrum mindset* to ensure that the work gets done and the team collaborates well. A Scrum mindset is the set of attitudes that individuals must hold such as respect, collaboration, willingness to learn, pride in ownership, focus on delivering value, and the ability to adapt to change. Discussion is presumed to be productive when everyone has a Scrum mindset. Scrum leaves the dynamics between the people on the team implicit. Scrum assumes both a well-functioning team and well-functioning individuals. This is often not the case and, as we have said, when we introduce diverse people, individual attitudes and biases may appear, especially through conversation.

Looking back to Ranya's experience and our intervention, we see an example of a workable approach. Rather than blaming the developer or anyone, we tune and clarify the roles and responsibilities of the group. Modulating individual behavior by tuning expectations within group practices is consistent with our approach to redesign practices to interrupt and redirect problematic behavior into something that advances the team's mission. This is the approach we take as we explore tuning Scrum. Let's start with the idea of mindset and self-management.

MINDSETS AND SELF-MANAGEMENT

Central to the Scrum mindset is self-management through discussion. But what does mindset really mean in practice? With a catchy word like "agile" (who wouldn't want to be agile?), it is easy for different people to project their personal assumptions and understanding on it. The best way to understand what people mean when they talk about an "Scrum mindset" is to translate it into specific observable behaviors. In other words, make it explicit.

Scrum principles champion the value of honest team communication to ensure the work gets done with quality. When everyone on the team is "like-me," communication may be easier. When the implicit value of straightforward, even argumentative interaction, while standing up for one's point is also shared and all members are listened to and valued equally, maybe it can work. But if the team is composed of very senior people, junior people, new hires, and maybe one woman—this is not a level playing field. Sylva tells us of her experience related to using mindset values in collaboration.

> I had another developer tell me, "Hey, you should be working on this ticket rather than sitting here chatting with Brent." I told him to mind his own business, but he just said, "Hey, we have openness as our value and I just wanted to be open about what I thought.
> Sylva, Software Developer

Clearly, Sylva and her colleague do not have the same idea of appropriate behavior for a value like "openness" or about who is entitled to criticize who. When most people on the team are "like-me," they often feel entitled to criticize people who are not "like-me."[31] Research also finds that people who see their authority threatened are more inclined to address harsh criticism towards someone who they see as a potential threat.[5] Sylva's colleague may or may not feel threatened but he feels entitled to criticize Sylva. Irrespective of the origin of this developer's action, he is undercutting Sylva's sense of value and belonging to the team. "Openness" that undercuts belonging and self-confidence also undercuts team cohesion and so productivity.

A central tenet of Scrum is to trust the team to work, plan, and *get the job done through conversation*. We do not agree that this so-called social dimension of Scrum is better for women—not when they are in a minority, not when they are in a role that has less power than others in the meeting, and not when they are new or early career.[20] It is not sufficient to put individuals together in a group, tag them "self-managed," and expect they will automatically know how to coordinate and work effectively as a Scrum team.

Unstructured discussion can open the door to bias and discourage participation

Worse, self-managed teams in Scrum explicitly rely on implicit processes, notably subtle control of individual team members by the rest of the team. The founders are quite open about this, calling it "control through peer pressure"[28] or "control by love."[26] From social psychology, we know that in self-managed teams, group pressure is always at the expense of minorities.[21] And in tech, women are a minority. For intersectional identities, e.g., women of color, the situation is even worse.[22]

We have observed that within self-managed teams, a vacuum can develop leaving no one responsible for ensuring fairness and making sure that group dynamics do not get out of hand. In teams that do not practice Agile, managers traditionally deal with interpersonal issues—but not in Scrum. Instead, "control by love," peer critique, and pressure is part of the espoused Scrum culture. The Scrum Master is the guardian of the process guiding the team in the "correct" practice. The Scrum Master is not empowered to stand up against something that the majority of the team would consider acceptable, expected behavior. A self-managed Scrum team has no guardian ensuring the kind of professional group dynamics that may be needed in diverse teams.

What if we tune the roles in Scrum? The *Gender Equality Report for Germany* includes recommendations on how to ensure gender equality in tech organizations. One recommendation is to establish a role to ensure gender equality in the team.[24] This Gender-Equality Master is a variation of Scrum Master. If this intervention is adopted, the team will have two coaches with potentially conflicting mandates. Another possibility is to redefine the role of the Scrum Master to include diversity facilitation. This change would require the Scrum Master to monitor and ensure equal participation from all team members and modulate highly critical behavior. This new role expectation is consistent with the Scrum Master's responsibility to ensure smooth collaboration.

But to be effective, the Scrum Master must be trained in diversity issues and have the skill to facilitate interpersonal interactions that enhance value and respect for all. This may work when the Scrum Master is in a central group working with multiple teams training and coaching them in Scrum techniques, facilitation techniques, and now effective diversity interventions. But if the Scrum Master is a member of the team who both works as a developer and performs the Scrum Master role, success may be more difficult. Will

Effective tuning builds on existing principles when defining new behavior expectations

they be able to stand up to more powerful members and/or the majority of the team? Will they be resented for performing this role? Will advocating for equality inadvertently communicate that women need "special" treatment, further ostracizing them? If the Scrum Master is a woman advocating for women, will it appear to be favoritism undermining team cohesion? For any effective role change of this kind, the whole team must agree that equal participation for all is a core team value. We are skeptical about tuning conversation by tuning the Scrum Master role this way without that agreement.

An alternative solution also consistent with Scrum principles is to redefine the team's mindset. This approach was developed in one of the Scrum teams we worked with. This group declared gender equality to be an explicit part of the team's mindset. Then they defined concrete actions which could be measured, calling upon the Scrum tenet of *empiricism*. This intervention helped the team move away from peer pressure and control by "love" as their technique to manage decisions. Michelle shared their redesigned approach.

Scrum's empirical approach has not only made our software better—it has also helped to modulate some team members' behaviors, for example regarding cherry-picking certain tasks. Our team consists of two women and four men, all with a strong commitment to gender equality. We have since started applying the Scrum pillars of transparency, inspection, and adaptation to our goal of reaching gender equality: We actually played around with metrics like the number of sticky notes per person, participation in meetings, happiness with the distribution of tasks, and who addresses problems; then we inspected the data and tried out changes. We changed the order in which we contribute to our meetings and have started to put tasks on the board that might otherwise go unnoticed (housework). Michelle, Solution Architect

Michelle's experience is indeed exceptional; many women we have talked to could not even imagine suggesting measuring things like airtime in their meetings. But Michelle's team shows how with commitment to gender equality, Scrum's tenet of empiricism can be used to create gender equality by design.[9] They collect data, experiment with solutions, and develop practices that nudge behavior toward equality. Combining mindset and an empirically driven approach allowed this team to minimize the undue influence and interpersonal pressure that might drown out the voices of early career people and anyone "not like-me." In other words, by introducing greater structure into the team's conversation practice, bias and unequal participation were interrupted.

Michelle helped her team pay attention to diversity issues by explicitly raising diversity issues as a new value within their mindset with clear behavioral expectations. This works because the whole team agrees to both the new value and the behavior change for the purpose of equality. With team agreement and a set of concrete interventions consistent with Scrum principles, a Scrum Master tasked with a diversity focus may be able to help their teams.

But not everyone will be comfortable or able to influence the team to change their mindset. Raising diversity as an explicit issue to address can work, but perhaps we can uncover other ways to tune conversation that is a sneaker way to attack the problem. If we can tune working meetings to improve a practice like decision making so it works for getting to decisions efficiently and helps women be heard, we do not have to designate it as a diversity approach. We sneak in more equality by tuning the practice. Let's look at an example of tuning Scrum's working meetings.

TUNING WORKING MEETINGS

Since Scrum is structured into a series of working meetings, we can step back and see how they are doing against our idea of how best to run a working meeting. In Chapter 8, we introduced what is needed for productive meetings using group critique sessions as an example. As we said, well-run meetings increase effectiveness.[15] The great thing about the working meetings in Scrum is that they are already highly structured, use physical artifacts to focus conversations, are time-boxed, and the Scrum Master facilitates. They also clearly define the mainline conversation—what is on or off topic. Each meeting has one central focus, each has a clear goal. With clarity around on and off-topic content, facilitation is easier. This structure effectively helps teams run the meeting well. Below we show how Scrum separates conversations by clearly defining what is on and off topic.

1. **Daily Stand-Up.** Mainline: Everyone answers the three questions and looks for blockers. Off topic: Fixing the blocker. This is a subsequent meeting. Even if it occurs with a sub-team right after the stand-up everyone knows it is a new meeting. It is explicit.

2. **Sprint Planning**. Mainline: Define sprint goal and figure out what to put in the next sprint. Off topic: Any discussion of additional possible stories or features, any discussion of things that happened in the last sprints.

3. **Review Meeting.** Mainline: Check the deliverable and how it did against the sprint goal. Off topic: Any evaluation of how well the process worked.

4. **Retrospective.** Mainline: What worked and didn't in the last sprint process. Off topic: Any discussion that rightly belongs to a different meeting.

Separating conversations is an important principle for facilitating team interactions and reducing interpersonal chaos during discussion. In The Critique Meeting, we separated providing feedback from resolving issues, clearly defining what content the discussion will address. Within typical meetings, team members often find themselves in conflict because they are talking about different topics in what appears to be a discussion—but they don't realize they are discussing different topics. For example, teams often argue about a particular feature, but

Separating conversations minimizes interpersonal conflict and bias

they are really arguing about different aspects of the feature. One person is extolling its value for improving the workflow, another hates the current graphical representation of the feature, and a third thinks the implementation is too difficult. The user researcher, designer, and developer all see the same function represented in a design differently.

To smooth the conversation the authors in their role as team coaches point out these different topics and introduce the idea of separating conversations. For example, we remind them that the first prototype is testing the need for the function at all. If users don't like it there is no need for us to worry about its visual design or development challenge. At this point in the design process, we remind them that the team is trying to define high-level functions so they can save the other issues for later. Once the team members can see the different topics, it is easier to get back on track.

Interpersonal conflict like this invites people to talk over each other; to overlook quiet, inexperienced, and diverse voices; and is more likely to promote nonprofessional behavior like yelling. These are all issues that women point to as undesirable aspects of technology's work culture. Open, unstructured discussions whether to make decisions, explore alternative perspectives, or generate creative ideas are the contexts where interpersonal chaos thrives. *Naming* a conversational topic or behavior helps the team and facilitators recognize them. For example, when two overlapping each separately. Naming, as we will see in the next chapter, is a good way to concretize a behavior or experience and so better manage interpersonal dynamics.

All conversation works best when the meeting itself is clear on what is on and off topic. Good meeting design defines what is in and out of focus and reduces potential conflict between

Conversation works best when all are clear about what is on and off topic

team members. Working meetings which already have tight goals like the Scrum meetings improve interpersonal interactions. But unfortunately, within their well-defined meetings Scrum practice still leaves many interactions unstructured. The emphasis in Scrum on decision-making through discussion opens the door to chaos and the implicit influence of bias.

Throughout the book, we shared research and stories of the impact of unmitigated stereotyping and bias. We pointed out that women may not feel welcomed or connected to the team; bias may dominate evaluation of their work and interactions; they may not be asked to do challenging work; and self-doubt and lack of support was a frequent experience. These experiences work against women easily speaking up in a group setting. Of course, we do not want to manage our teams by ensuring they never talk to each other. We would lose the dynamic nature of the interactions that women so value. But we can help improve conversation by tuning the structure of working meetings to ensure that equal participation works. As an example, let's explore Planning Poker which we discussed briefly in the Introduction to the book.

TUNING PLANNING POKER

Planning Poker is a structured conversation to estimate the "cost" of a story to develop. Cost is typically estimated as story points or developer days. Story points are a unit that is influenced by the amount of work, complexity, uncertainty, and other factors. With story points, a value of cost in time and effort is assigned to each story or task. The actual numbers do not matter, only the relative value of one task when compared to another. For example, a task with two story points is twice as big as one with one point; a task with eight story points is four times as big as the one with two points. Planning Poker gets its name because after creating private estimates members reveal their score—like in a poker game.

The typical process looks like this. Estimates are first made privately, then revealed to the team. If all the estimates for a particular story are essentially the same, that result is accepted and the conversation is over. Because of silent estimation, Planning Poker already ensures that women and early career people form and express their opinion. But if the team does not agree they use conversation to determine what estimate to accept. Here is our opportunity to tune the process so that everyone gets an equal voice. The issues we need to address include the following:

- Do we ask each individual to defend their estimate? Do we implicitly devalue the opinion of women or less experienced developers who just entered the field? Do very senior people speak so authoritatively that we all just capitulate to them?

- From a quality perspective, should we listen more to people who are usually right? Or have more experience? If so, how do people learn?

In the current process, people with high estimates and low estimates justify their estimate to the group. Then discussion continues to close the gap. This requires people whose thinking is not in-line with the majority to argue for their ideas. Under these circumstances, it is easy to give in to the power of majority pressure—both for people who are unsure of themselves but also for people that are truly experts.[19] Early career and less experienced members may feel put on the spot and fear getting it wrong. We observed that these team members focus more on guessing what the rest of the team will estimate rather than giving their honest estimate. Quiet team members and women, or other diverse people who may not have a strong

> *Diverse perspectives are lost if people don't share ideas that are not in-line with the group*

sense of belonging or confidence, may also hesitate to share their estimate. But if everyone doesn't speak for their estimate, the team cannot hear their reasoning which might influence the group to create a better estimate. The diverse perspectives are lost.

Interestingly, expert team members are also reluctant to share. When they come up with an estimate that is widely different from those of the rest of the group, they worry about appearing arrogant. To preserve group harmony, they may give up, especially if they aren't responsible for or likely to be working on that story. But when the estimate agreed to does not reflect the difficulty of the task, the team will not achieve their goal. Less experienced individuals who may be responsible for that story are set up for failure. Again, if the team does not hear the reasoning of the experts, success is handicapped.

Open discussion to resolve an estimate of story length, does not work for anyone. And it certainly does not invite everyone into the conversation. Here are some ways to tune this meeting that we have tried that may work for you.

Comparable stories. After each sprint, the team assesses how well they did on their estimates in the review meeting. Using the principle of empiricism, the team now has real data of how long certain stories took the team to do. It can also allow teams to develop reference stories. Tune discussion about proposed estimates by asking people to explain their estimates based on stories that are similar to those the team has completed in the past. For example, "This story is like the story in the last sprint because...so I estimate it to be the same as it actually took." We have found that people can make reliable estimates using a very simple three-point rubric: The work is a little bigger than, same as, or smaller than an example story. Using a rubric with comparable stories gives the team a language to explain estimates in a way that everyone can understand.

Criteria of good estimation. Estimation by comparable stories helps the team start to characterize stories by various criteria that are naturally used by expert estimators. Criteria of good estimation work the same as criteria of goodness in The Critique Meeting; they provide an agreed-upon language for why the estimate is reasonable. If the expert estimators on the team help derive these criteria, their knowledge is built right into the team's decision-making process. Now the team can use the same rule of engagement as in The Critique Meeting, stating that every person must provide a "why" with an estimate.

But sometimes the "why" of an estimation is influenced by underlying bias. For example, often the stated estimates assume who will be working on the story. This combines perception of the skill of that individual with the difficulty of the task. This kind of estimate can hide assumptions that women are less competent than men. Or the team may assume that the most skilled person will always do the challenging projects limiting who gets to take on a challenge. One way to make these assumptions explicit is to ask for example, "How long will the more skilled person take?" and "How long will someone else take?"[4] If there are differences based on who will do the work, the team can alter estimates to support growth in team members and a more equal distribution of challenges.

Also, bias about the task can impact estimation. When team housework is put on the board it may be estimated as having less cost because it is a devalued activity, even though it may take more time. So, ask the person who last took notes how long it took and use that to inform a more realistic estimate. Exploring estimates in this way can derive new estimation criteria or surface new rules of engagement for the meeting.

Build in estimation training. Criteria of good estimation create a training technique for new hires and early career professionals. Work Buddies can introduce the criteria using example stories from the last couple of sprints. The team can support and encourage developing estimation skill with feedback, perhaps in an estimation review meeting (like weekly design critiques). Since growing early career people is a team goal, in Planning Poker ask early career professionals to share their estimation first, followed by more experienced developers. These practices support both growing skill in estimation and listening to every estimate with an eye toward honing the criteria of good estimation.

With these changes, we redesign Planning Poker so that everyone's thinking is both heard and encouraged. (See tuning planning poker process.) We change the focus of the estimation discussion from defending one's position to explicating their criteria for the estimate. We explore how we think about estimation through questions to the estimator to get at the root of the estimate. As a result, the team gets more information for decision making. After everyone shares, the team votes silently again. Hopefully, now a shared understanding is reached and the estimation is over. If not, the team accepts the estimate of the majority. If the team is just learning to work together or has decided to tune their process, add a quick Process Check (see Chapter 11) to reflect on how the estimation process went right after the planning meeting while everyone remembers how it went.

These suggestions add structure to the discussion in the meeting, which levels the playing field for women and those with different levels of skill. These suggestions also help the team improve the *velocity* of their sprints. Velocity is the average amount of work that gets done in a sprint and is an important measure of the productivity of a team in Scrum. Being good at estimation and getting everybody's input is essential for completing the planned stories within the sprint timeframe. Velocity is an indicator of a successful Scrum team. Improving estimation in these ways helps the team succeed; it is simply good practice, but it also sneaks in ways to address issues of diversity.

Making the implicit explicit by adding structure to discussions as we have done with these Scrum meetings can be tuned for any working meeting: Define a clear goal, the mainline conversation of what is on and off topic; define a clear procedure for the meeting; establish criteria of goodness relevant to the conversation; structure interactions so all participants interact and share their thoughts productively; and build in training and growth opportunity. Tuning working meetings helps address team effectiveness, which matters for

> ### Tuned Planning Poker Process
>
> 1. Estimate individually and silently to allow for diverse ideas.
> 2. Share estimates in round-robin style, with less experienced people going first to encourage their participation.
> 3. Everybody explains their estimate, not only team members whose estimates diverge.
> 4. Reference criteria of good estimation or comparable stories when explaining your estimate.
> 5. Provide a why, or new criteria, for any estimation. Grow team skill at estimation.
> 6. Estimate silently a second time after discussion. Start the round robin with a different person than in the first round.
> 7. Choose the estimate of the majority.

team success. But tuning working meeting also helps women and diverse people participate, which builds team cohesion and creativity. Tune working meetings by building on the existing principles and practices of that type of meeting, as we have done with these Scrum meetings. Sneaking in changes this way is good for making diverse teams work—and achieving the team's mission.

Last, let's turn to another Scrum process, picking stories. Picking stories are a key issue for women to ensure that they get challenging and stimulating work.

PICKING STORIES: CHOOSING WORK

The culmination of the planning meeting is choosing what to work on in the next sprint. Scrum places a high value on individual team members choosing their own work, the stories they will work on. But does that work for women, early career professionals, and others who are underrepresented on the team?

Scrum methods, particularly Scrum, encourage team members to choose work for themselves. This is considered one of the hallmark practices of empowered Scrum teams. This practice is referred to as self-assigning, signing up, or pulling. Self-assignment is not implemented as regularly as other Scrum practices, such as daily stand-ups or iteration planning.[18] Instead, stories may be assigned to a specific person in some implicit way,[4] for example, by assuming "This is front-end, so Thomas will be doing that." Unfortunately, choosing stories does not take place in a meeting that can be structured—it is an ongoing process throughout the sprint. And not all the tasks the team needs to do are on the planning board, notably team housework.

Who gets (or rather takes) which tasks is crucial to how successful any individual is seen by the team and within the organization. We have already discussed how women tend to take on and are expected to do team housework, which does not enhance career goals.[13] In Chapter 2, we discussed how important stimulating work is to women and the barriers and biases women face in getting challenging technical work. And we have noted in Chapter 3, sometimes women are hesitant to take on a challenge, often related to gender expectations. So, they may need The Push and Support. Challenging work is at the center of retaining women in tech—and so picking stories is a target for intervention.

The way individuals choose stories in Scrum can let in this bias. A "grab a task" approach or asking, "Who wants to do what?" allows the most confident and verbally dominant people to get what they want to work on through the power of their character or influence. And unlike women, men are more likely to be given tasks that enhance their career and allow them to "make their mark." If women are then assigned lesser tasks, they can't "make their mark." As a result, women wrongly appear to be poorer performers.[16]

Small changes can make a process work better for diverse teams—and everyone

Taken together, our challenge in tuning story selection is to ensure that work challenge is distributed more equally.

Below, we share ways to tune task assignment in Scrum. Try out some of these suggestions. Brainstorm with your team better ways to ensure that challenging work, housekeeping, and boring work can be more equally distributed.

- **Measure your process.** Use an empirical approach. Watch out for consistent patterns of task pulling. Often, certain team members repeatedly take certain types of tasks. This might be due to their expertise, their preference, their status, or because they have always done it. Identify patterns and decide whether they are okay for everyone or whether you want to change them.

- **Define criteria of goodness.** Make this your basis for distribution of tasks and monitor how you are doing in Retrospective meetings to reflect on what is working for the team.

- **Promote others.** Build appreciation and The Push into your process. Encourage team members to suggest another team member work on a story, especially an early career team member. Explain your suggestion by referring to the criteria of goodness and describe why you think this person should do the task, emphasizing their value.

- **Round robin.** This is always a good and simple way to distribute challenging work. Try a random number generator to determine who grabs a story first or next. Or create a fixed order for getting stories or tasks so challenges and housework fall to everyone some of the time.

- **Early-career first.** Let new hires and early career professionals pick first as a commitment to growing their professionalism. Don't keep them from picking the hard stuff; in daily stand-ups they can declare blockers and get help from a more senior buddy. If your velocity seems to be affected, do it less frequently or let them buddy with an experienced person to learn. But don't relegate less experienced people or diverse people you "think" are not as skilled to doing boring work—they will leave.

- **Challenging work vs. boring work.** Identify what stories are cool or challenging and which are more like housework. Make sure all the work that needs to be done is actually on the board. If a team member gets a boring story last sprint, let them pick a challenging one this time around and vice versa. Then no one has to ask for challenging work—it comes as part of the process. And boring work is equally distributed.

- **Be guided by rules of engagement.** Try our rules of engagement for picking stories and develop your own. (See rules of engagement for choosing work.)

The goal of picking stories is not just getting the work done. It is also to ensure you develop the skill of early career people and distribute stimulating work for all—including women. To design these interventions, we again make small changes to the existing process. Equally distributing challenging and boring work makes the team responsible for all of the types of tasks. This equity

> ## Choosing Work:
> ## Rules of Engagement
>
> 1. Make sure all the work that needs to be done is on the board.
> 2. Everyone works on challenging and boring work equally.
> 3. We support growing the skill of new hires and early career professionals by making sure they get challenging work.
> 4. Don't be a "hog" looking for all the best work.
> 5. Offer help and support if someone is hesitant to take on challenging work.
> 6. Make appreciative suggestions to others about which tasks they could work on.

builds team cohesion and sends a message of value to each team member. If everyone feels that they are getting challenges and that their skills are recognized at the right level, you will retain women and build skilled professionals. Tuning processes this way, making practices explicit and equitable, is simply the best way to ensure product excellence. That it also improves the collaboration of diverse teams is our sneaky side effect.

THE POWER OF EXPLICIT PRACTICES

Redesigning key processes—onboarding, critique, Scrum—is one of the best ways to interrupt bias and improve daily work life for women in tech. Raising awareness of gender issues in workshops is not effective because awareness does not translate easily into behavior during everyday practices. But redesign that focuses on making implicit expectations, practices, and values to make them explicit, sharable, and able to be monitored can make a huge difference to the success of diverse teams and women's experience.

Sneak attacks on key processes for diverse teams call for finding small changes that can be easily adopted. Wholesale changes to processes are hard to accept and to adopt. The Analysis Matrix helps identify important targets for change to processes. Then you can redesign the implicit

Find small changes that can be easily adopted that improve practices of diverse teams

aspects of the practice to sneak in small changes that help everyone on the team be more successful but also address issues important to women and other diverse populations. Build these changes upon existing principles and into existing procedures. as we did with suggested changes to Scrum. We do not advocate eliminating Scrum and replacing it; we advocate iterating it with an eye toward diverse issues and making the whole team more successful.

Declaring that these changes are to increase equality when the whole team has not chosen to redesign for equity, suggests that women and other diverse populations need special treatment. Special changes for underrepresented people in tech further ostracize them. It suggests they can't perform up to expected standards. But if the whole team is willing to explicitly adopt a commitment to diversity and equal participation, all the better. Use the Analysis Matrix in a workshop and engage everyone in redesigning how they work.

Scrum has some very good practices, but it can be made better for product quality and issues that impact women. Unfortunately, any implicit practice has the potential to undermine women's success and experience of belonging to the team. Implicit expectations may reduce women's participation, introduce bias, and undervalue the contributions of women. Indeed, we have discovered that even explicit practices like self-managed teams and reliance on freewheeling discussion for critical decisions can breed biased practices. But even these practices, as we have suggested, can become more explicit and so interrupt this bias.

In our discussion of possible interventions for Scrum, we have shared both our thinking about possible solutions and the recommended changes. To recap, we have shared ways for you to equalize participation in discussions, distribute challenging and boring work equally, and use measurement to track against the process the team adopts. We have emphasized the need for clear structure and procedure in working meetings as well as the power of defining criteria of goodness and rules of engagement. Try them out with your Scrum teams and adapt them to any team practice.

Small changes like those we have suggested can increase team cohesion and participation by all members of a diverse team. When that happens, you will start to reap the dynamism that fuels innovation. Improving the work practices for women and diverse teams improves the way the whole team works—and so the success of the work product. Tuning practices with small changes make them more likely to be adopted. That these changes also benefit women and diverse teams make them sneaky. We will never make everything in a process completely explicit. But using the Analysis Matrix can help redesign how you work to interrupt bias and improve women's experience of daily work.

> *Sneak attacks through small changes can increase team cohesion, participation, and innovation*

Tuning key processes can make a huge difference to women's experience at work—and to retention. But not everything happens in working meetings which can be managed. To retain women in tech we must also address interpersonal dynamics, as we do in the next chapter.

CHAPTER 10

Valuing and Jerk Behaviors

In the last three chapters, we offered interventions to interrupt bias and manage interpersonal chaos within some of the core processes in tech work. As we said, interpersonal chaos can end up excluding and devaluing underrepresented populations and those with less experience. In Part I describing the @Work Experience factors, we have also emphasized the centrality of interactions with the team, manager, and co-workers for women to thrive—and for retention. They are the source of a sense of belonging, value, coaching, and stimulating work. Managing interactions between co-workers in the workplace is key to improving team cohesion and creativity. Interventions into working meetings can booth reduce interpersonal friction and disrupt differential treatment of underrepresented team members.

Working meeting structure will not always eliminate interpersonal interactions that create friction or offend. Unfortunately, interpersonal issues are an often-cited reason women do not experience belonging or feel valued. Women report a significantly more negative workplace experience than men.[6] They may not feel interpersonally safe when they participate in meetings.[5] In the context of the technology industry, women report feeling alienated when the company is what they call a "bro" culture.[2, 4, 8] The everyday parlance is that men can be "jerks."

But which exact behaviors communicate value, and which exact behaviors communicate devalue or worse are seen as "jerk" behavior? If we knew, could we design a more valuing team culture? Could we more easily modulate jerk behavior? Could we have direct conversations about these key behaviors impacting the experience of daily work life?

To change behavior, start by knowing the exact behaviors you want more or less of in the workplace

The Valuing and Jerk Project was created to better understand workplace interactions in order to identify specific behaviors that create or undermine the experience of connection and value. To identify the behaviors, we conducted 26 field interviews with women and men, 25–45 years old, working in a wide variety of jobs and technology industries. The interviews focused on finding observable behaviors that stimulated the feeling of "value" and those that represented being a "jerk" to the participants. Interviewees were asked to describe incidents of valuing and jerk behavior they experienced, witnessed, or did over the last month. We collected incidents from working meetings, manager and co-worker interactions, and non-work life contexts. The field data was interpreted and organized into key issues and behaviors in an affinity diagram using Contextual Design techniques.

This qualitative data identified 15 key valuing and 15 key jerk behaviors. It also revealed that both women and men experienced, witnessed, and sometimes engaged in both behaviors. To better understand the impact of these behaviors, we next developed a survey and ran it with over 100 worldwide participants, male and female, from typical job types and industries in technology. The survey asked people to share the importance of these behaviors to them. We were interested in whether some of these behaviors are generally more important than others at a population level. By comparing answers of men and women we were able to explore their differing experiences.

We found that these behaviors resonated with survey participants, and that some of these behaviors are more important than others. For the tech environment we studied, we also found some behaviors enhance or diminish value differently for men and women. We do not claim that we have identified every behavior that someone may respond to, but we do argue that our research has defined key themes and behaviors we can work on. The three key themes are briefly described below:

My Skill and Job Role. People pay attention to how their skill or their job type is valued by co-workers in their team, in working meetings, or by their manager or higher-level managers. When their contribution is acknowledged and celebrated, they feel valued. When it is ignored, they experience feeling devalued. When co-workers actively devalue and denigrate their skill, those people are considered to be jerks.

Making a Relationship. People experience value and devalue based on whether others create relationships with them. When co-workers, particularly those with greater status, invest time in a person by coaching, supporting, and listening, it communicates value. The person feels known and that they matter to the organization. But if collaborators ignore emails or requests, don't look at or get to know a person's work, or don't listen or listen impatiently, they are seen as jerks. This jerk behavior is experienced as denial of a relationship and personal worth.

Violating Expected Behavior. This last group of behaviors relates to the definition of expected professional behavior. People feel valued when others are clear about work expectations and generally treat co-workers as professionals. Professional behavior includes communicating work status in a timely way when, for example, their work is going to be late, or they are going to be late for a group meeting. But when co-workers violate these behaviors, they are jerks. Yelling, using foul or inappropriate language, or disrespecting others' life commitments by being consistently late is considered jerk behavior.

To help teams talk about the valuing and jerk behaviors we created The Valuing and Jerk Character Posters. Each poster has an evocative name and uses an animal to represent it. The characters help make a set of related behaviors salient—and are fun.

The poster intervention uses the principle of *naming* we introduced in Chapter 9. By giving a set of behaviors a name and persona, we help teams and managers see the behaviors that matter to people. With this awareness, individuals can self-identify what to work on and teams can decide what behaviors they want to promote and exclude in their team culture. Naming and personifying the behaviors to make them real is the first step in change.

Below, we describe our six valuing characters (The Honorables) and our six devaluing characters (The Jerks). We also highlight any differences we found between men and women. We later discuss ways of using the posters in a workshop and in conversations either one-on-one or with the team. Download all the related materials here.

THE HONORABLES

The Honorables represent the valuing characters and the behaviors people experience as the most

valuing. They are presented in the order of most important based on our survey. But note that the Thank-You Maven, Consummate Professional, and Grateful Maker are roughly equal in importance.

The Coach is our top valuing character. The Coach is a more experienced person who helps, guides, or partners on hard projects or problems. They take time to listen to and understand my work or personal needs. They invite questions and give straightforward feedback calmly with respect and a desire to help. They help identify my strengths and use them to forward our career. The importance of

The Coach is not surprising given our discussion of Local Role Models in Chapter 4. The Coach is willing to create those valuing relationships we need in the workplace.

The Coach is important for both men and women. But women's scores are significantly higher than men's. Given the challenges women face in the tech workplace, The Coach behaviors are essential. Taking time to coach validates women's value in the workplace.

The Reward Master recognizes my skill tangibly through promotion, bonuses, or other material forms of value. The Reward Master stands in second position because getting a promotion is direct career-enhancing recognition of skill or other contributions to the team or organization. Money and formal recognition are always a straightforward statement of value.

Other aspects of The Reward Master are spot bonuses from colleagues, extra bonuses for especially well-done jobs, or trips and conferences. But if these extra recognitions do not turn into promotions they fall flat.

Tangible value, especially promotion, is seen as very important for men and women but is the top valuing behavior for men. Promotion is much more important than spot bonuses or other tangible nods to accomplishment for everyone. From a valuing perspective, promotion denotes value recognized by everyone in the organization.

The Thank-You Maven consistently delivers direct straightforward verbal appreciation of my skill or behavior. They may tell me directly after meetings or they may send an email detailing the value of my contribution. Acknowledging the good contributions or participation of co-workers seems to be built into the Thank-you Maven's character.

This direct value is even more powerful when the acknowledgment is in public, at a group meeting, for example, or if they tell senior management about my good work or how well I manage my team. Simple thank-yous and statements of value are the easiest behavior we can increase. Both men and women appreciate appreciation. Understanding this, some companies now send a weekly reminder to take a moment to acknowledge teammates.

The Consummate Professional represents people who hold themselves to expected standards of professionalism. The Consummate Professional embodies an overall attitude of respect and consideration expected from managers and co-workers in a healthy workplace. For example, The Consummate Professional communicates their status or need for help in a timely way so others can adjust. They deliver work on time or let the right people know so that the team's goals can be achieved. They come to meetings prepared and on time. Others' time and home commitments are a top consideration for a Consummate Professional. Respect for time and commitments is valued by men and women alike.

Also, The Consummate Professional communicates work expectations clearly so employees can successfully fulfill them. Lack of clarity can lead to negative feedback or rework when leaders fail to lay out their goals and criteria for success. This is particularly important for women who are asked to demonstrate their skill over and over—unclear expectations for work undermine women's ability to succeed.

Overall, The Consummate Professional hold themselves to a standard of professionalism that emphasizes the impact of their own communications and behavior on others.

The Grateful Maker is a co-worker or team member who uses and desires my work, my skill, and my overall contribution. The Grateful Maker acknowledges my contributions to a project or a working meeting. They may build on my work or ideas with acknowledgment. Using what a person produces with acknowledgment is one of the greatest statements of value.

The Grateful Maker also delivers value when they ask for my participation or leadership in an upcoming project or meeting or when I'm given a challenging project without asking for it. Requesting my skill, asking for me, is a direct statement of value. It is the way we know that we matter to the team.

Grateful Maker behavior is important to both men and women, but it is one of the top aspects of value for women. Acknowledgment of a contribution to the team's work is a statement of women's contribution to the mission, reaffirming their skill and importance to the team.

The Champion advocates for my ideas, career, or issues that I need to address. The Champion is looking out for my success and career growth. They may arrange to showcase my work to senior management to help build my presence. They ask me to speak publicly and help me stretch my skills. They support me when I want to take risks by giving me air cover to upper management.

Overall, The Champion helps me think about my career growth and how to build relationships and position myself for advancement. They "have my back" and are "in my corner." We often call The Champion our mentor.

The Champion is important to both men and women, but it is more important for women. Unfortunately, in tech companies women need advocates to be sure their ideas are heard and work is seen—and for promotion.

Explicit valuing matters to everyone. And it is essential to new hires and fostering successful professionals. But for women trying to succeed in tech culture, valuing is critical. In our survey, of the 15 valuing behaviors, 6 scored above a 4 on a 1–5 scale indicating their importance for women. In contrast, the top scores for men were all below 4. As we have said, the technology culture and society come with attitudes and expectations related to women that disadvantage them. Valuing behavior that counteracts this bias is especially important for women. Taking the time to coach women to success, promotion to recognize their skill and contribution, acknowledgment of their work, advocating for them, and simple thank-yous are all ways of raising up women's value to the organization. And valuing builds women's personal power. Valuing behavior is important to men, but valuing behavior is critical for women to thrive.

THE JERKS

These characters represent the jerk behaviors we want to discourage or deliberately manage. They are also presented in the order of importance based on our survey. You can see that in all cases they undermine the kind of value that The Honorables represent. Note that regardless of order, The Values Violator and Decision Dominator are about equal in creating a negative experience.

The Megaphone Mocker is our most hated character. The Megaphone Mocker yells at co-workers in public. The Megaphone Mocker loudly criticizes others, declares shortcomings in their work or the irrelevance of their job type, or disparages their personal style. The Megaphone Mocker may use inappropriate or foul language when referring to co-workers. The Megaphone Mocker yells in disparaging ways in meetings, one-on-ones, and the hallway. This character can humiliate a person or a team. Unfortunately, The Megaphone Mocker often has more power and influence than the co-workers they are disparaging.

Directed yelling of any kind is the worst jerk behavior for everyone, but it is more important for women. Yelling can feel like an attack when directed at an individual person. Since the tech culture is frequently described as one that rewards yelling and bullying,[3] to retain women in tech we need to address The Megaphone Mocker.

The Demeaner tells me my work, ideas, or job role are not worth consideration or are substandard. Demeaners don't take the time to understand what I produce or do, so they don't really know. The Demeaner shows up most in one-on-one interactions.

When The Demeaner is a more senior person important to my career, they undercut my self-confidence and success. They may criticize my work but do not help me improve it. Or they may redo what I produced without providing feedback or collaborating. If they listen to what I share, they listen impatiently. The Demeaner's words and attitude feel like an uninformed dismissal of my value.

Demeaner behavior comes in second as the worst behavior for women. For men, this jerk behavior is less important, probably because their competence is not questioned the same way as women in a tech environment. Men are not confronted with the necessity to prove themselves again and again. This might allow men to dismiss The Demeaner in a way that is not possible for women.

The Values Violator breaks the rules of expected behavior. The Values Violator represents dislike of people's behavior when they act as though social norms or corporate policies simply don't apply to them. This includes any implicit or explicit norms of the team. People see The Values Violator when someone does not abide by articulated organizational rules and expectations. These include policies of the organization, implicit expectations for collaboration within a team, and norms guiding professional work and behavior.[17]

In addition, The Values Violator also reminds us to think about the origin of our different expectations. People from different geographic, family, organizational, or team cultures will have different ideas of what constitutes a values violation. What seems like bad behavior to one person may not be perceived that way by another. The Values Violator tells us that people may perceive jerk behavior because of their individual experiences, challenging us to clarify expectations of acceptable behavior particularly at the level of the team. The Values Violator is disliked equally by men and women.

---

[17] Other research also shows that breaking typical norms in the workplace is considered rude and impolite, showing a lack of regard for others.[7] Breaking norms is also linked with lower retention, health issues, and reduced job satisfaction.[1]

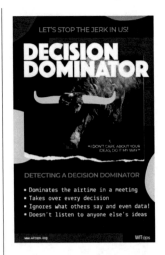

The Decision Dominator shows up in meetings. When the meeting is called to make a decision or generate ideas, The Decision Dominator is often a person of influence. They insist on their own way and will not listen to anyone else's point of view. They ignore input from others, known data, or best practices. The Decision Dominator takes up all the airtime, makes all decisions, and determines the direction of a project. This attitude communicates straightforward dismissal of anyone else's skill or knowledge.

The behavior of The Decision Dominator is disliked similarly by men and women, but women strongly dislike it when people in meetings insist on their own way and do not listen. The Decision Dominator may be an underlying culprit in the well-researched finding and complaint by women of being interrupted. The underlying attitude of The Decision Dominator is that others' ideas or points of view have little worth. If women, other underrepresented people, or new hires are perceived as having less worth, others with greater status may steamroll them like The Decision Dominator does. Our recommendations for well-structured meetings can interrupt Decision Dominator behavior.

The Time Waster can be a manager, direct report, or co-worker. The Time Waster is just that: they do not think about how their behavior impacts others that they work with. The Time Waster does not communicate their work status or the need for help, often resulting in the whole project being late. They come late for meetings and do not let anyone know. They may finally show up but are unprepared. Their work is late. They implicitly deny collaboration by ignoring email and requests from co-workers.

Managers can be Time Wasters too. They can ignore emails and requests from direct reports. They may ask for work to be done but then ignore or not use it. They may forget that they asked for the work. And worse for women, they may not communicate expectations for work clearly, causing work to be redone. At best The Time Waster is rude, at worst they communicate that other people's time commitments do not matter. Time Waster behaviors are high on the list of jerk behaviors for men, but less so for women. Men really hate when their emails, calls, and requests are ignored.

The Role Stomper is our least jerky set of behaviors. But role stomping can get in the way of smooth interactions between co-workers. Each job role comes with a set of expectations that include what work they do, decisions they can make, and meetings they should be in. The Role Stomper does not have someone's job role but acts as though they do. They make decisions, take over responsibilities, and finalize recommendations that are not rightfully theirs to make. The Role Stomper takes over my job and cuts me out of the work.

The Role Stomper may also go to key meetings that I should be in; they fail to tell others to invite me. The Role Stomper may be the person who used to have my role or another co-worker who seems to be taking over my job. Sometimes role stomping happens through miscommunication; two collaborators who fail to cleanly separate who is supposed to do what can feel angry when both do work that the other thought was theirs. Whenever role stomping happens, The Role Stomper undercuts my feelings of belonging, self-confidence, and the value of my job. Good managers pay attention to the problem of role stomping.

No one likes jerk behavior, not women and not men. Both women and men list yelling as the top jerk behavior—but it is rated even higher for women. We don't need much more evidence than this to challenge companies and teams to excise yelling from their team culture. Yelling is just an extreme version of disdain for another person and their value to the team, but it is not the only jerk behavior that is so denigrating. The Demeaner, The Decision Dominator, and even The Role Stomper all attack the value of our skill and work. When women's competence in a tech culture is already underestimated these kinds of behaviors directly undercut value and self-confidence. Our other two jerk characters represent unprofessional behavior; together, they denote what it looks like not to be a Consummate Professional. Jerk behaviors are not just a women's issue—all of the jerk behaviors matter to men as well. For a healthy and thriving company and team culture, we need to address them. Let's look at some ideas.

INTERVENTIONS IDEAS USING THE VALUING AND JERK CHARACTERS

The Valuing and Jerk Characters are fun and compelling. They draw attention to the issues of how we communicate with each other. This allows us to build discussions around them. Hang the posters on real or virtual walls to keep them top of mind. Below, we describe how managers can use the posters to raise awareness and foster discussion. Start with a presentation of the Valuing and Jerk Characters with silent reflection by participants. Then follow up with discussion soon afterward using group or individual meetings. Plan the best way to have these potentially difficult discussions. Try out our suggestions for a group or an individual approach.

Self-recognition of our own behavior issues helps us change

PLAN YOUR APPROACH

You may be leading a product or design team, or you may manage a group of people that do not work together on a project. Use your knowledge of your group to decide how public or private you want your discussion and interventions to be. When planning a group discussion, consider the following issues. You may be able to talk about valuing behavior as a group but need a more protected interaction for jerk behavior.

- If you have a trusted relationship with your group, you may facilitate the discussion yourself. Or, invite a facilitator so you can fully participate, and to help manage any friction.

- If your group is cohesive and collaborates well, you may feel comfortable with a group exercise for both valuing and jerk behaviors. Even well-functioning teams benefit from finding out how they can improve their interactions. If your group already had friction, start with valuing behavior as a team. A focus on valuing may be a good first step to smoothing relationships.

- If you think the group or individuals may feel exposed and unwilling to share, you may opt for a more anonymous approach like using a digital whiteboard that hides names.

- If you do not think a team approach will work, you may choose to follow an awareness presentation with discussions in your one-on-one meetings with participants.

INTRODUCE THE GOAL OF THE EXERCISE

Start by introducing the exercise to your group. Frame it as a fun way to understand individual experiences and improve the overall way the team collaborates. Emphasize that the purpose of looking at these characters is so each person can identify what kind of valuing or jerk behavior is important to them personally. Remind everyone that nobody is perfect; we all can do a better job.

Emphasize that the purpose of the exercise is to focus on yourself, not identify good or bad behavior in others. Pointing fingers and blaming others for issues in the team will most likely stimulate anger and defensiveness. Others commenting on our behavior when unasked is rarely welcome, especially with jerk behaviors. Even publicly saying that someone is a great Thank-You Maven, for example, by exclusion, implies that others in the group are not. Labeling another as a character just because we show one of the behaviors some of the time also can inappropriately pigeonhole them. Labeling doesn't allow for individual and circumstantial differences that may be associated with one incident of undesired behavior. Explicitly discourage team members from calling out individuals on their jerk behavior.

Focus on your own behavior—not the flaws in others

On the other hand, *self-recognition* and talking about the behaviors within individual and team conversations can help. With the characters as a stimulus, individuals may have realizations about their own behavior: "I'm doing a little of that behavior, but I need to do more or less of that behavior." Self-recognition is a key principle of change; self-recognition allows each individual to see where they want to change themselves. For example, one manager was able to admit that he was always late for meetings after he learned about The Time Waster. Then he found that issues of time emerged in his team conversation. By committing to change, he improved his behavior.

Reflecting on the valuing and jerk characters also helps the team assess how they are doing as a team. Then they can brainstorm ways to do better. Below we describe how to run a conversation to explore

Managers must commit to behavior change too

the team's experiences of valuing and jerk behaviors. Reflecting on valuing and jerk characters may impact the team's code of conduct. In Chapter 11 we will introduce the Team Manifesto and the Process Check as a way to help the team manage what is working and not working in how they work and treat each other.

PRESENT THE CHARACTERS AND REFLECT INDIVIDUALLY

The first step to facilitate discussion of valuing and devaluing in your group is to present the Valuing and Jerk Characters. Then follow up with individual or group discussions soon after. Present the characters yourself or watch our video as a group to begin.

During the presentation, individual participants silently reflect on their own experiences. Participants fill out the scorecards shown below to record their thoughts. Present The Honorables first while participants reflect and fill out the scorecard. Present and reflect on the Jerks second. Following presentation and silent reflection, hold discussions with participants according to your plan.

VALUING SCORECARD				
THE HONORABLE BEHAVIORS	Mark Valuing behaviors you received in the **last** month.	Mark Valuing behaviors you should do more of.	Mark Valuing behavior your team could do **more** of to improve team culture. Note the situations.	Mark Valuing behaviors that are most important to you. What are **your** top 3?
The Coach				
The Reward Master				
The Thank-you Maven				
The Consummate Professional				
The Grateful Maker				
The Champion				

JERK SCORECARD				
THE JERK BEHAVIORS	Mark Jerk behavior **you have received or observed** in the last month.	Mark Jerk behaviors **you have done.** What do you want to change in yourself.	Mark Jerk behavior **your team seems to find acceptable** that should be changed. Note the situations.	Mark Jerk behaviors you dislike the most. **What are your top 3?**
Megaphone Mocker				
Demeaner				
Values Violator				
Decision Dominator				
Time Waster				
Role Stomper				

Tuning behavior, just like tuning processes, is an effective way to encourage or interrupt behavior. Do not dictate ways to change behavior. Instead, brainstorm tiny interventions that people are willing to do to help increase valuing and modulate devaluing behaviors in themselves. In planning, you may opt for individual or group discussion, or a combination. Below, we describe ideas for how to run these discussions.

INDIVIDUAL DISCUSSIONS WITH MANAGERS

If you have opted for individual conversations, ask your direct report which valuing behaviors they find most important to them and discuss how to make sure they get more of it. For jerk behaviors, ask if they have identified behaviors they would like to improve. Also, ask for feedback for yourself for both the valuing and the jerk behaviors. Follow the feedback rules: Take it in and be grateful for it; do not discuss, do not justify. If someone identified a behavior in themselves, brainstorm ways to help them pay attention to or to change the behavior. If they are willing, suggest they ask a trusted individual to be a buddy who will support their change goal.

On the valuing side, a person may realize that they are not making themselves available to really coach or that they aren't responding to requests. Using their trusted buddy, they can identify one early career person who has requested help and start responding. If they are simply not acknowledging their teammate's efforts enough, they can commit to sending one valuing email or text a day or a week. Try using our valuing emojis and stickers in the download for the posters.

For jerk behaviors, explore solutions together that can help the person interrupt their own undesirable interactions. For example, one person with Decision Dominator tendencies asked a colleague to put up a stop hand signal when he had gone on long enough. In remote meetings, this can be a chat message. Another individual committed to writing down all her ideas that she feared the team would not consider and checking if the content was covered at the end of the meeting. She discovered that often they got to it. A third person, known to take up all the airtime, limited themselves to talking two times in a meeting. Tiny interventions like these are good ways to help people self-modulate behaviors that are getting in the way of teamwork. If they can change their behavior for a few weeks, it often sticks.

TEAM DISCUSSION FOR VALUING

Start your team intervention with a discussion of valuing. Post The Honorables posters or a combined poster. After the presentation, use the Valuing workshop template below to collect ideas, organize, and brainstorm ideas for change. Ask the team to put up sticky notes on a real or virtual whiteboard. Each person records the name of their top three valuing characters on three sticky notes. These are the characters most important to them that make them feel most valued. You may use a different color sticky note to denote the valuing characters they think the team needs more of.

Valuing Characters	What top 3 behaviors are most important to you?	What top 3 behaviors should the team do more of?	Brainstorm solutions and values
The Coach Reward Master Consummate Professional Grateful Maker Thank-you Maven The Champion			

30 minute Process: Identify – Sort – Brainstorm – Choose

Next, sort the responses to see both what matters to individuals and what needs to be increased in the team. Lay them out to see what the top valuing characters are for the group. Discuss people's thoughts on the top characters to identify the most important behaviors the team should focus on for change. Brainstorm how to increase these valuing behaviors within the group and commit to a plan.

Try out the changes for a few weeks and assess how you are doing and whether you should iterate your approach to becoming more valuing. Use Process Check (Chapter 11) to continuously monitor and tune how you are doing as a valuing team.

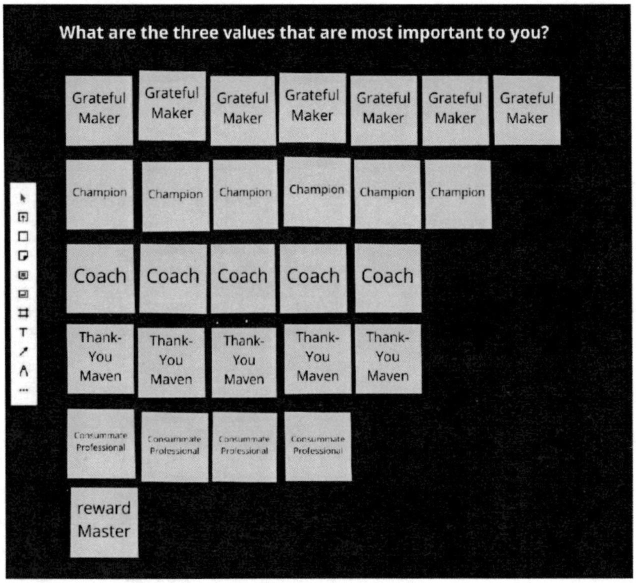

Figure 10.1: Results of one team's Valuing Workshop showing the most desired valued behaviors.

Talking about desired valuing behaviors using the posters helps the team express how they feel—as long as it is in a safe environment. Putting desired behaviors on a board (Figure 10.1) externalizes individual feelings while reflecting the desire for more valuing behaviors in group interactions.

Seeing the votes of others in the team allows for individual reflection. For example, the manager of one group we worked with learned that the team needed more acknowledgment for their work. By seeing The Grateful Maker votes from their workshop at the top of the board, the manager and team members could talk about how much they valued and would like more acknowledgment for their work. This manager did not realize that people wanted more and explicit recognition of their work. Team members were able to share that they felt offended when someone uses their work without giving them credit by name. During the discussion, the manager was also able to hear who shared and who did not. They resolved to follow up with quiet people in their one-on-one meetings.

Using a simple process like this helps get the issues out. The posters frame the conversation much better than just asking a group, "How can we be more valuing?" The presentation allows people to examine their own experience and be more prepared to talk; it encourages them to resolve to increase their own valuing behavior.

If you want more anonymity, set up a virtual whiteboard that treats everyone as a guest so no one can see their name. A physical whiteboard where real sticky notes are posted is also more likely to hide individual responses as everyone writes their notes at their place and then posts all at once. Remember, collecting ideas without discussion, or through simultaneous input on a virtual whiteboard, invites voices from quiet or underrepresented team members such as women.

Valuing behavior is easier to talk about than jerk behavior. But a discussion on valuing can implicitly call for a reduction in jerk behavior. Many jerk behaviors are the flip side of valuing. For example, a discussion of what people are looking for from The Consummate Professional can result in identifying being on time as an issue. They are referencing The Time Waster, without any finger pointing. So always run the discussion about valuing first.

TEAM DISCUSSION FOR JERK

A direct discussion of jerk behavior uses the same process we described for valuing behaviors, focused on the jerk characters. Introduce the discussion following the presentation as a way to understand what the team collectively needs to work on, not to target or blame individuals. Emphasize that all of us can engage in these behaviors some of the time. Your goal is to find the most disliked negative behaviors to work on as a group. Pretending to discuss

jerk behavior as a whole when people are really targeting a single individual is not productive. So be careful the discussion does not turn into blaming. If people have self-recognized their negative behaviors during the presentation, they can be more comfortable working on change within the context of expected team behavior.

Use the Jerk Workshop template to guide your team to identify jerk behaviors they would like to address. Then brainstorm and organize solutions in the brainstorm section of the template. Again,

Design sneak attacks to foster valuing and decrease jerk behaviors

solutions do not target individuals; they address changes that can interrupt negative behavior within the group's practices. For example, the group may change their code of conduct, facilitation, practices for participation, and other sneak attacks on their processes. Once the team identifies jerk behaviors they want to monitor or change, start by imagining the positive change that would occur if the jerk behavior was stopped.

Consider this example. Different people may have different ideas of problem behaviors represented by The Values Violater. One person might want everyone to greet each other in a friendly way at the beginning of a meeting. Another might feel the team wasn't helping each other enough on hard problems. A third person might think that the team wasn't using respectful language as they understand it to be. Often for the Values Violator, we find that team members have conflicting ideas of social norms particularly when they come from different cultures. Once the team sees that The Values Violator is a top jerk behavior for the team, they can do a round-robin to share the positive behavior they would like to see instead—without targeting any individual. (See the What is Rude? exercise in Chapter 11 for ideas of how to run this discussion.)

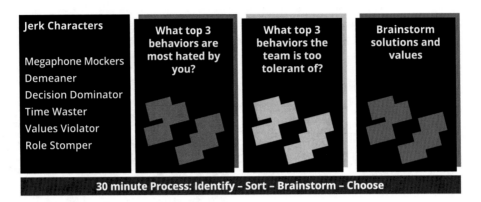

Jerk Characters	What top 3 behaviors are most hated by you?	What top 3 behaviors the team is too tolerant of?	Brainstorm solutions and values
Megaphone Mockers Demeaner Decision Dominator Time Waster Values Violator Role Stomper			

30 minute Process: Identify – Sort – Brainstorm – Choose

Once the group is clear about the positive behaviors they are looking for, they can brainstorm techniques to modulate behaviors during working meetings and team interactions. For example, the team might agree to spend a minute in friendly conversation at the start of a meeting or generate a better way for people to indicate when they need help. Perhaps, the team can create rules of engagement and empower the facilitator to intervene when an unwanted behavior occurs. Or, they decide to use a timer for participation to be sure no one dominates a meeting. All of these changes are sneak attacks to interrupt negative behavior that appear in working meetings.

Try the intervention for two or three working meetings, and then talk about it. But be careful not to target one person. Again, if the team discussion is really an attempt to change the behavior of one person, a coach or manager needs to step in to help that person make a plan for growth. Whether jerk behaviors are talked about in team or one-on-one discussions, be sure everyone is ready to have the discussion. Presenting the jerk characters before conversations raises the issues without targeting individuals; talking about behavior in the abstract diffuses blame and helps the team define acceptable behavior for the team.

Remember, all group discussions, whether valuing or jerk, can be followed up with individual discussions in one-on-ones with managers. But to have these discussions you need to build trusted relationships with your own team—and be willing to be honest about and change your own behavior. Then everyone can create personal growth goals to hold themselves accountable to as we discuss in the next section.

BECOMING A VALUING ORGANIZATION

Interpersonal dynamics are at the center of creating a Dynamic, Valuing Team. How we treat each other affects our feelings of being valued and the team's overall cohesion. If we do not feel valued by our team and manager, it undercuts our ability to contribute with confidence. If we do not feel valued, it also undercuts our feeling of belonging to the group. If we do not have coaches, champions, and Local Role Models to work with, we do not feel valued. And if people are impatient or ignore our work, we will feel dismissed. If all the people on diverse teams do not feel a sense of value, we will not get their best. And since feeling valued is something that women and diverse team members experience too little in tech, we need to focus on ensuring all people feel valued to retain them.

The behaviors identified by the Valuing and Jerk Project tell us exactly which behaviors communicate value and exactly which behaviors undermine it. These may not be all the ways that value and devalue are experienced, but these are certainly a good starting place. Everyone can be more valuing. Everyone is a jerk some of the time even if they didn't intend it. So, everyone can examine their own behaviors and make a commitment as to how to be more valuing. Similarly, everyone can examine their behaviors and resolve to or get help to reduce their behaviors that communicate devalue.

We encourage each individual, including managers, to identify *personal growth goals* whether it is part of a formal review or not. Managers can discuss these goals in one-on-ones but more important is that individuals choose and commit to their own goals. Personal growth goals do not need to be something a manager oversees. Instead, find a trusted buddy to support you and tune your own behavior. To increase valuing behavior, you may need to create reminders to be valuing. If you are a leader, you may commit to coach or champion someone and make a

Define personal growth goals for yourself and hold yourself accountable

plan with scheduled interactions. To become more professional, start by clarifying what professional means. Leaders can engage the team in defining what professional means for them to develop a shared understanding. Define what it means to be a professional; include giving others credit for their work. Write a note on your desk or laptop with the three valuing behaviors you want to be reminded of. Hold yourselves accountable for increasing your own valuing behavior.

For jerk behavior, define a substitute behavior to replace your habitual behavior, such as modulating your dominance of the airtime by writing down thoughts or limiting your participation. If you catch yourself listening impatiently, can you diagnose why you are doing that? If you are time-compressed, reschedule the meeting so you can listen. Do you devalue the work of an individual? Why is that? If being late is your habit but the team defines professionalism by being on time, put in more reminders. Hold yourself to your own scheduling commitments or change them. Talk things over with your trusted buddy. Ask them to let you know when you are violating your own participation commitment. Write a note on your desk or laptop with the top jerk behaviors you want to be reminded of, and the behavior you will do instead.

But what about our most serious jerk behaviors The Megaphone Mocker and The Demeaner? Some companies explicitly include "no yelling" in their corporate code of conduct. But this only matters if it is enforced. To be clear, we are not talking about the occasional raised voice because people are upset, frustrated, or stressed. We are not talking about the emphatic way people talk when they come from cultures in which that is the usual tone of expression. We are talking about directed yelling: yelling at a person or a team while emphasizing their worthlessness. Being human, people may yell some of the time as our data has clearly shown. But any such

Apology is the most powerful way to counteract your own jerk behavior

outburst requires an explicit apology. Apology can go a long way to healing a negative interaction if the person commits to managing their outbursts so they do not happen again. But if outbursts are frequent and an acceptable part of the culture, if outbursts by people with greater power and influence are deemed "something we have to live with," we are creating a hostile working environment. Yelling at people is simply not acceptable behavior in the workplace.

The Demeaner is the second-worst jerk behavior for women. Demeaning their skill and work product is not going to help women feel a sense of belonging. Demeaning skill is not going to help new hires and early career professionals grow into successful, contributing members of the team. To be clear, Demeaner behavior is not about getting feedback that your work product is not perfect; nor is it a less than enthusiastic response to your contributions. Getting feedback is simply hard, which is why we addressed how to do it well in Chapter 8. Effective feedback comes with an underlying message of valuing and a commitment to the growth of the professional. True Demeaner behavior comes with a strong devaluing message; it is filled with meanness and direct dismissal of another person's worth. If we find that we were too harsh, we catch ourselves and apologize. We commit to improving our own interactions with others. But if a work culture tolerates personal denigration embodied in The Demeaner and Decision Dominator behavior, it also creates a hostile working environment.

On diverse teams, individual differences and styles abound, all of which impact interpersonal interactions. The tone of a working environment is not solely the result of individual behavior. A well-known truth about teams is that the team is not the sum of the skills of the individual. When any two or more people collaborate, communicate, or interact what happens between them depends on how well all these underlying differences mesh, which alliances exist, what history the team has, which organizational culture it is embedded in, and many other factors. Successful teams, partnerships, and management relationships are not only about individual behavior; they are about the mix between the people. So, managing valuing and devaluing behavior at the level of the team makes sense. Engaging the group in brainstorming how to improve how they interact raises up the behaviors that matter for that particular group so that they can choose how they want to interact.

Everyone and every group can do better if they want to. To create a valuing culture, we must become aware of behaviors that matter, determine what we as individuals and teams need and should work on, and then hold ourselves accountable for our personal and group change. The culture of the company and the team may explicitly and implicitly expect and condone particular behaviors. A corporate code of conduct that is buried in the intranet or hung on a wall disconnected from daily practices has no teeth. How leaders model valuing or let jerk behavior slide sets up implicit standards for behavior. How meetings are run can encourage valuing behavior or tolerate jerk behavior. To make change in how we interact, teams must take responsibility for explicitly creating a valuing team culture. They can create a culture in which diverse teams thrive.

For retention, become a more valuing culture

In this chapter, we have identified valuing and jerk behaviors that matter to people working in technology and on technology teams. We have made them explicit so that we can design ways of interacting that encourage or modulate the behaviors on diverse teams. But one workshop and a discussion on valuing and jerk behavior are not enough. To make team values real they must play a part in everyday team processes. In Chapter 11, we share techniques to help teams build and monitor their commitments to be a well-functioning and valuing culture.

CHAPTER 11

Building Resilience: Team Manifesto and Process Checks

In Chapter 10, we called upon teams and individuals to hold themselves accountable for their behaviors in order to create a more valuing culture. In Part II overall, we provided recommendations and tools to tune practices so they work well for women and everyone. In this chapter, we describe interventions to help teams explicitly define the values and behavioral expectations that will guide how they work and collaborate. But values are only real if the team's work practices and behaviors reflect these values. The team must hold themselves accountable to their own standards. Through ongoing reflection, the team can monitor and tune their work practices and interactions to ensure work success and professionalism.

The Team Manifesto is a tool to help teams articulate their team values, how they want to work, and expected professional behaviors. The procedure to create the manifesto also includes raising awareness of interpersonal differences, equal treatment, professional behavior, and home/work balance. With a defined set of values and expectations, the team can hold themselves accountable to uphold them. But accountability only comes through continuous reflection and tuning of their practices. The Process Check is a technique to help the team reflect on how well they are doing in the work and against their values.

Deliberately create a positive team culture— and reflect to stay on track

Through it, the team can identify practice glitches, interpersonal chaos, and problems getting the work done professionally. Taken together these two techniques form a coherent practice to help teams become more resilient, cohesive, productive, and valuing.

Research on resilient teams, both remote and face-to-face, tells us that resilient teams are reflective teams.[4] People on resilient teams recognize they can react to mundane everyday happenings by attributing negative intentions to team members, for example, Sue is lazy, unreliable, and without skill. These attributions create a cycle of devaluing. But people on resilient teams can also question their reactions. Through discussion, they rethink their negative attribution and find solutions that improve how the team works. Perhaps Sue was late because she needs help with a hard problem— not that she is unreliable. In other words, the resilient team looks at breakdowns in the team's processes and interactions instead of blaming or denigrating a person. They tune their perspective on the situation, which leads to a change in practice. Teams that cut each other some slack and focus on solutions are more successful.[14] And teams that support each other build team resilience.[1]

Naturally occurring resilient teams use a reflective process to enable them to continuously improve how they work together. But reflection works better if teams articulate their values and expectations first. The Team Manifesto and Process Check taken together help build success and resilience. We have used these tools with many teams. They make team culture explicit and physical so that teams can self-monitor and improve themselves. We recommend these two processes for any product team, management group, or set of people who are expected to work together for an extended period of time. Both practices can easily be used by remote teams by capturing information in a collaborative document or online whiteboard.

THE TEAM MANIFESTO: DEFINING TEAM VALUES

The Team Manifesto is a charter, a shared agreement of how to work together. It is the team's statement of the commitments they will judge themselves and their success against. These commitments are developed by the team members to outline team-specific values and norms regarding teamwork and professionalism (e.g., attendance, airtime, participation). Team performance benefits from an explicit discussion about values and standards for working together.[7] Also, teams with a Team Manifesto are better at handling turbulences, which in turn increases performance.[15]

Team values and expectations can only be monitored if they have associated behaviors that are observable. A good Team Manifesto contains explicit values with explicit associated behaviors

Every team member co-creates and commits to the team values

to which every member of the team commits, including the manager. It establishes shared expectations for the team members and team behavior. Only through voluntary commitment to a co-created document can we expect people to modulate their own behavior. Then if team members do not live up to the Team Manifesto, the team can revisit their values and norms. Through reflection and tuning, the team can address negative behavior and prevent exclusion,[9] as well as improve their work practices.

There are different approaches to developing the norms and values for your team.[2, 8, 12] What's important is that the approach used allows everyone to have a voice in creating the kind of team they want to be. The approach we describe here has been used by both of us, and we have adapted it for different teams.

The typical Team Manifesto process uses a stimulation artifact like a word cloud to get the team thinking about what kind of team the group wants to be. We also use a word cloud tuned with an eye toward issues related to diverse teams. But to help teams get to know each other's differences we recommend that the team first do an interactive exercise exploring their perceptions of rudeness. (See the What is Rude? exercise.) The exercise raises awareness of expectations for professional behavior before creating the manifesto. It is a fun icebreaker for new teams too.

We cannot dictate the team's values and expected behaviors, but we can set the context for reflection before creating the manifesto. Then team members will be more likely to consider acceptable and unacceptable behaviors, along with behaviors that ensure productivity and creativity. In this way, we build in issues of importance to women.

Below, is our description of the overall process for creating a Team Manifesto.

STEP 1: RUN THE EXERCISE EXPLORING RUDENESS

The rudeness exercise works well to help teams think about issues that can affect diverse teams and produce interpersonal conflict. Interestingly, when asked this question people tend to identify behaviors that are related to valuing and jerk behavior. Asking about rudeness is less interpersonally difficult than asking directly about valuing and jerk behavior, so we prefer it.[18] A simple question about the meaning of "rude" allows people to share their perspective, which can be informed by their culture, personal annoyances, and past ways they have been treated. As such, it raises up expectations that may get in the team's way.

What is Rude Exercise

Do this exercise before you create a Team Manifesto. It is also great when you start a new team, merge teams, or add a person to the team.

Time: 30-60 minutes

Use physical sticky notes or a collaborative whiteboard.

1. Each person answers the question "What is Rude" by writing their perception on a sticky note. Make as many sticky notes as you like recording one perception per note.

2. Put all notes in a central pile.

3. Sort the notes into similar behaviors. Team members can now see where their perceptions overlap and where there are differences.

4. Discuss the differences and learn about each other's perspective.

5. Build your Team Manifesto, including agreed-upon professional behavior

STEP 2: BUILD A TEAM MANIFESTO

The Team Manifesto is a document created by the team that answers the question, "What kind of team do you want to be?" Display a visual stimulus like the word cloud below. We have developed this word cloud to include things that are relevant to issues that women may have.

You may instead write stimulating questions on a flipchart or digital document. We have also used sets of questions such as: "How should we give feedback;" "What do we feel about being on time"; "What do we want to do for home/work balance"; etc. This helps the team reflect on the kind of team they want to create.

The word cloud or questions orient the team to values and expectations that affect work quality and how they work, aspects of collaboration, and interpersonal relationships. Both the rude-

---

[18] Save an explicit conversation on valuing and jerk behavior for when the team is up and running and interpersonal conflict is getting in the way.

ness exercise and the word cloud increase everyone's sensitivity to issues of being a productive valuing diverse team. Together they encourage the team to build in values and behaviors that work for everyone on the team. Follow the steps outlined below to create the manifesto. (See the process for building the Team Manifesto.) See the example of the manifesto created by the team that created the Career Power Boardgame.[19]

Process for Building the Team Manifesto

Preparation:

- Complete the rudeness exercise first.
- Use physical sticky notes on a wall or a collaborative tool like a whiteboard which mimics sticky notes.
- Hang the word cloud on the real or virtual wall for thought stimulation.

Goal:

The team answers the question "What kind of team do you want to be?"

Procedure:

1. Thinking of the word cloud stimulus, write your answers on a sticky note. One answer per sticky note.
2. Team members read their words round-robin style collecting like notes from others. Or put all notes in a central pile.
3. Sort notes into piles naming the behavior or value each pile represents.
4. Choose the values and behaviors the team wants to commit to.
5. Write the value name on the left and the natural opposite on the right. Include behaviors that represent the values.
6. Cluster and prioritize the value themes. Select 5–8 key values by voting and discussing until the list is narrowed.
7. Record the Team Manifesto on a flipchart or document.
8. Sign the poster to show your commitment.
9. Get your managers to sign to gain their agreement too.

---

[19] Career Power is a board game designed to help professionals in technology discuss real work challenges while having fun. Learn more https://www.witops.org/career-power-the-board-game/.

The Team Manifesto is easy to build and can then be used for reflections. It helps teams define the culture of work and relationships that they want to be held accountable for. Display it during every Process Check or Agile Retrospective. Be honest about how you are doing.

You may find you need to add a new agreed-upon value. Update your Team Manifesto as needed. If you need to merge two teams who each have a Team Manifesto, they can look at their manifestos to see what they already agree on and tune changes to form a new one. If a new person joins a team, be sure to reexamine the manifesto as a group. When you get a new hire, you may want to do the rudeness exercise again and regenerate the manifesto. Remember that each person must agree to the values and behaviors if they are going to be accountable to them. Listening to the voice of each new person also communicates that they are a valued member right from the start.

Game Team's Team Manifesto	
POSITIVE VALUE	**NATURAL OPPOSITE**
1 • Work Ethic • Use Best Practices • Show up on time • Respond to Communications • Do assigned tasks	**1 • Lazy** • Making up designs based on personal opinion. • Not answering emails • Not completing tasks • Wasting time
2 • Have Fun • Put effort into team building • Take breaks • Know each other • Respect life commitments	**2 • Be Strangers** • Expect team cohesion to just happen • No breaks • Getting angry if everyone doesn't agree with you
3 • Work for Excellence • Be excellent at every stage • Plan for failure • Make something real	**3 • Build for a Grade** • Do minimum required • Procrastinate • Make something just so-so
4 • Be United • Support each other • Own the results • Straight talk, no whispering against each other • Kill your personal babies	**4 • Be Divided** • Put each other down • Disagree in public (outside of the team) • Talk about people behind their back • Cling to your ideas that don't make sense
Signatures	

The Team Manifesto is your explicit statement of the kind of team you want to be. It references how you want to work, values around what you will produce, and what a professional relationship looks like. The Team Manifesto is the creation of the team culture. It may not embody all aspects of your culture like any socializing traditions as defined in The Team Onboarding Checklist. But it is the description of norms, expectations, and acceptable behavior for everyone on the team in the context of getting the work done.

The local team culture has the greatest impact on issues of inclusion, exclusion, and valuing. Even if the company as a whole is less welcoming, the local team culture can create a haven for diverse teams. Change at the local level is very powerful. One creative and productive team with good practices can influence other teams and trickle up to management. But a Team Manifesto only works if it guides all team behavior. To ensure a living commitment, you need continuous reflection and improvement. Only through continuous tuning of how you work together can you catch problems both in the work and between people. But our goal in reflection is to improve our processes, how we work, not blame people.

TUNE PROCESSES—DON'T BLAME PEOPLE

Before the pandemic, we still had remote Retrospectives because our teams were in three differ-
ent places although a small group of people was in each location. We used voice, not video, for
these meetings. I hated Retrospectives. Everyone saved up everything they didn't like until the
end of the sprint and then listed all their complaints, making them personal. I just put my mic
on mute and cried! Now we use Retrium.[20] The tool makes everyone enter what went well and
what didn't, with a suggestion for improvement—no complaining. Now our remote meetings
are more productive and civilized. And because I now have a leadership role, if someone starts
blaming someone else, I just cut them off. I remind them that we are here to fix processes, not
gripe about others. I tell them to handle interpersonal issues during the project, not at the end.
Ginny, Product Owner

At the heart of our approach to change is to focus on how we work together. The Team Manifesto along with continuous reflection help the team pay attention to improving their practices and interactions. In Scrum the Retrospective meeting is meant for reflection. The Process Check, shared

Focus on changing
principles and practices—
not personalities

below, is a reflection process for other types of teams. But without vigilance, reflection meetings can devolve into blame and accusation. In Ginny's story, meeting chaos occurs because the Retrospective procedure is not articulated, expected behaviors are unclear, there are no rules of engagement nor a moderator. Introducing the Scrum Retrospective tool helped because it also introduced structure to run the meeting and collect observations. If we want to get the benefits of reflective practices, we must be sure that they, like all other practices, are well structured.

In Chapter 8, we shared how difficult it can be to get feedback on a work product. It is also difficult to get feedback on how the team is working. Reflecting on how we work is essential to create resilient high-performing teams. But feedback is more challenging for a new or less cohesive team and those with many inexperienced people.[5] And too often, as in Ginny's case, people may see a Retrospective as a gripe session.

People are not used to focusing on how they work versus what is wrong with each other's work. Process is invisible—personality is not. Blame, accusation, and telling another person to behave differently is not a "fix" to the process—and it undermines the feeling of value, belonging, and cohesion between team members. Unfortunately, Scrum may inadvertently encourage blame. Scrum guidance for the Retrospective meeting describes the kind of content the team can address in a Retrospective: "Scrum Team inspects how the last Sprint went with regards to individuals, interactions, processes, tools, and their Definition of Done".[11] The guidelines explicitly state that the team should inspect individual behavior in this public context.

---

20   Retrium is collaborative tool for running remote Retrospectives. There are many tools specifically dedicated to retrospectives. Other teams use templates on their physical or remote whiteboards for their Retrospectives.

Scrum encourages teams to evaluate each other, not just the way we work. It focuses on evaluating others, not on evaluating ourselves and our own participation. This focus will always lead to blame and negative critique; it does not help the team to work together, and it undermines team cohesion. And unfortunately *inspecting individuals*, as Scrum calls it, comes with biases.

Throughout Part I we have shared attitudes and expectations in society and in the tech culture that disadvantage women. Even in the most gender-equal countries, there is strong evidence that all people evaluate men and women differently.[10] Feedback to men is typically more specific; women receive vaguer and less actionable feedback.[3] Feedback for men is actionable and direct—less so for women. Sylva shares her experience.

> *When I started on this team, my teammates told me I should feel free to approach them if I needed help with anything. I could tell that everyone was well-meaning and sincerely wanted to be there for me. But when I was unsure about a piece of code or if something was not clear, I would try to figure it out by myself. There were only a few times when I was really stuck and would ask one of my team members for help—it just seemed like a large threshold to interrupt their work to help me. Then when Ben joined the team, he was told that if he had a problem or was unsure about something, he should just add a feature branch in the repository system and file a pull request so that the others could see the question right next to the relevant comment in the system. They actually guided Ben the way to get very specific feedback on the code right where he needed it. With me, they had just told me to ask for help if I needed it. Of course, I am also using pull requests in the repository system now, but I do wish they would have told me about that right away. Sylva, Software Developer*

Sylva was told to ask for help. But for Ben help was built into the process of coding. He did not have to "bother" an experienced person; he had an experienced buddy to work with. When anyone makes an error, we must remember that they are always working within an implicit or explicit system of support and skill development. Sylva was not set up to succeed because the team did not provide her with the same system of support as Ben. Blaming Sylva for poor code in a Process Check conceals the different ways men and women are treated. The solution is not to "fix" Sylva, the solution is to change the process of support.

Also, women, as we have said, are both perceived and treated differently than men. Men are more often praised for their assertiveness, while women receive more comments on being "aggressive". [13] Women are more likely to be praised for having—or criticized for not having—communal traits. Men are more likely to be judged on their expertise[10] and told straight up where they need to improve. Women are less likely to get straight talk about their performance, instead, they are told "white lies," that the quality of their work is okay when really there are areas that need improvement.[6] Again, when we look at individual performance, we need to

Give women specific coaching that improves their work—just like men

look at the processes we use for evaluation—and our underlying assumptions. Before blaming anyone for their participation or performance, the team needs to first examine if each member of the team is receiving the same constructive feedback and coaching. The Team Manifesto can commit to providing constructive feedback and support to all as a value. Then the Process Check examines how the team is doing in providing skill support.

Inspecting individuals as a goal in a Retrospective does not serve women—or anyone—well. This is why the Process Check rules of engagement disallow blame and focus the team on improving practices. But as we recommended in Chapter 10, people can declare their own growth goals and ask for help during a reflection. Then all can acknowledge their improvement as it happens or plan better support. And teams can commit to equity and interrupting bias in their practices. They can devise ways of measuring or tracking their own success. Then the reflections can examine how the team is doing against these commitments. If you find differences for men and women or other typical biases that you want to focus on, further tune your practices, rules of engagement, or values.

Reflection helps the team keep their work and collaboration on track

Reflection is essential to ensuring that teams live up to their Team Manifesto and also work together smoothly to produce successful products. They are essential practices if we ensure that their rules of engagement and procedures work to interrupt bias and blame while tuning how the team works together. We share the Process Check procedure next which can be used by any team and in Retrospective meetings as well.

THE PROCESS CHECK

A Process Check is a simple technique for evaluating how the team is doing in its daily work and how well it is following its manifesto. Whether you use a flipchart, a collaborative document, or an Agile Retrospective tool, the team captures what works and what doesn't. The Process Check is run similarly to The Critique Meeting (Chapter 8) or a Retrospective in Scrum. The moderator and rules of engagement ensure the meeting is productive and does not devolve into blame.

The Process Check might happen every few days in the life of a brand-new team, but run a Process Check a couple of times a month for ongoing teams. For Agile teams with long sprints, waiting until the end of the sprint might be too long. Introduce Process Checks or add 10 minutes for it to a stand-up every week or two. Process Checks work with any kind of team: UX, development, cross-functional project teams, etc. Any time people work together closely to produce an outcome, a Process Check helps keep them on track. Build it into your practice. When you feel that you are not managing yourselves well, run a Process Check.

RUNNING A PROCESS CHECK

To run a Process Check, the team starts with the positives in a round-robin format collecting at least one positive from each person. Remember this is a good time to recognize each other's contributions or their progress on a personal growth commitment. Next, the team uses the same procedure to list negatives. (See Process Check example) For negatives, we do not publicly call out perceived failings or skill deficits in others. This, as we learned, is jerk behavior. Blaming others for bad performance is explicitly disallowed. Instead, we look for process issues that may have contributed as we discuss below.

While listing the negatives, participants privately generate design ideas to improve the negatives. When the negatives are complete, everyone puts up their design ideas. The scribe records everything as it is shared. There is no discussion or evaluation of any comment, although team members may say they agree with what is already there. Then the team walks through the ideas for improvement and elaborates on solutions collecting thoughts round-robin style. Last, the team votes or gets some form of verbal agreement to try some ideas out in the next timeframe. Use a moderator to ensure effective participation just like in The Critique Meeting.

Teams may run a longer Process Check at the end of a project. If a team is doing regular Process Checks, the last Process Check looks more like a discussion of key processes and values. It is also a reflection on how the Process Checks may be tuned to work better for the team. This last Process Check is also an opportunity to give each other kudos for work done well.

> ### Process Check Example
>
> **Pluses: What Works**
>
> + We are starting meetings on time
> + We have running code!
> + We all went out for lunch together.
> + We ran a critique, and everyone listened. It was productive.
> + The customers reacted positively to the prototype.
> + I (Nika) feel confident about my design and thank Michelle for her help.
>
> **Minuses: What Doesn't**
>
> – We didn't give ourselves enough time for this part of the work.
> – Our stand-ups are too long.
> – We are wasting time arguing about word definitions.
> – We are not getting feedback on our work fast enough.
> – I'm (Joe) still worried I'm talking too much in meetings.

Process Checks work best when team members are face-to-face in a meeting or remote with videos on. This way all can hear and see each other, including non-verbal reactions. As people share, team members get a flavor for how everyone is experiencing the work. If a Retrospective tool or template is used, all members can put in their thoughts before the meeting, but the members' contributions should be shared in the meeting so that together they can discuss issues and choose the next step. All team members must participate and add positives and negatives. With or without a tool, discussion must happen. Discussion works if you moderate well and use techniques to be sure everyone is heard. Be inspired by the techniques we have suggested.

See the description of the procedure and rules of engagement that we use below. Tune them for your own use. The key is to encourage the team to own the process and improve it.

Process Check: Procedure

The following procedure can be run in-person or remote. If simultaneous input is used, follow this procedure to share what is being written real time.

Length
- 10-minute stand-up add-on
- 30 minutes once a week or bi-weekly
- 60-90 minutes at the end of project, sprint, etc.

Roles
- Team Members share what works and what doesn't for the timeframe reviewed.
- The scribe captures notes onto a flipchart or in a collaborative document that all can see.
- The moderator ensures participants follow the procedure and rules of engagement.

Reference Rules of Engagement for positives and negatives defined

Step 1: Finding what works and what doesn't
1. The scribe starts the process by asking what worked in the previous timeframe.
2. Team members state one item in turn until all positives are recorded.
3. Move from person to person in a round robin way. The scribe and moderator get a turn.
4. Repeat for what doesn't work.
5. Anyone who has a design idea to improve the process writes it on a sticky note.
6. At the end, all put their idea on the minus it addresses.

Step 2: Choosing what to try in the next timeframe
7. The scribe reads off all design ideas for each item.
8. Team members discuss which ideas to try with moderation of the discussion.
   a. Team members may ask the advocate of the design idea to describe it in more depth.
   b. Discussion may generate new design ideas which may work better to address the issue.
9. After exploration, the team votes on which design idea to try out.
   a. Try voting by marking the desired ideas simultaneously rather than public hand raising.
10. Team tries the new ideas for the next timeframe and evaluates again.
11. If new values are uncovered the team decides to add something to the Team Manifesto or holds a separate meeting revisiting their commitments.

Process Check: Rules of Engagement

Attitude

- Improve the procedures used to get the work done.
- Be sure support or resource issues are addressed and improved.
- Focus on improving process not blaming people.
- All contributions are welcome; none are not heard.

Start with positives about what works

- If you have negative comments, you must first state a how the team worked well in this round.
- Be specific about what practice, meeting, or interaction worked well. .
- Positives include:
  - Processes that worked well, particularly if it is something new that the team tried.
  - Adherence to the values and principles of the Team Manifesto.
  - Effective collaboration and interactions.
  - Accomplishments of the team and individuals: doing good work, helping others, or reaching a self-defined stretch goal.

List negatives, while generating design ideas

- Be specific about what practice, meeting, or interaction did not work well.
- Negatives include:
  - Elements of a process or a problem in the work impeding progress
  - Violating the values and principles of the Team Manifesto
  - Ineffective collaboration or work interactions
  - Instances, not people, that might be exclusionary or where not everyone was heard. Do not name names.
  - Need for additional help or resources
  - An individual declaring that they personally did not perform as they wanted to (implicitly asking for help)

No discussion during list-making

- Round-robin to collect comments.
- Use a process to solicit contributions from all.

Design idea discussion focused on process improvement

- Explicit ideas to change the process to address the problem raised
- Acceptance of the outcome of the group

Listen to the moderator

- All agree to allow the moderator to modulate participation according to these rules of engagement and the meeting procedure.

The best way to stop interpersonal irritation is to keep the work process productive. Being responsible for improving the way you work is also best at the local level. Even if an organization embraces a particular practice, the team can iterate it to meet the needs of their team members. They can optimize their practices for their specific project, team member skills, and life constraints. If someone needs help, they can declare it in a Process Check. If a team has committed to Non-judgmental Flexibility, if someone has extra demands from home, the Process Check is a good place to declare and solve it. If the negatives listed in a Process Check start to look like issues of valuing or jerk behavior, the team may need to be reminded of their overarching value to be valuing and professional. By reminding the team of their commitments as a group, without singling out an individual, we refocus everyone on improving collaboration.

Stop interpersonal friction by keeping the work productive

The Process Check is the team's tool to help themselves work professionally. Build it into your core practices. If your team seems to be devolving into chaos ask yourself when was the last time you ran a Process Check?

CREATING A RESILIENT TEAM

A team always creates its own culture, rules of engagement, roles, and practices. They implicitly define how they will work through their day-to-day interactions. But culture that is implicitly formed is prone to the pitfalls of bias and the influence of the most powerful members of the team. And if the team's practices are not well defined, the lack of clear roles and structure invites interpersonal chaos.

Defining a Team Manifesto makes the values and expectations for good teamwork explicit. It can explicitly build in equal treatment for all along with ways to measure it. Then in Process Checks, the team can reflect upon and improve the way they work. They can examine if they are manifesting their values; if they need to tune their processes; if they are collaborating well and professionally; and if they are falling into the pitfalls of bias. Everyone on the team participates in creating the manifesto and in reflecting on how they work. In this way, the team owns co-creating the work culture they live in. By tuning these practices to include reflection on issues that matter for women, we ensure the team discusses them. Taken together, we help create resilient teams.

The Team Manifesto and Process Checks are pieces in the puzzle of how best to retain women in tech. But explicit values and reflection are not enough to ensure a cohesive, dynamic, and valuing diverse team. Teams are not natural process designers. They are focused on getting the work done. Nor are they interpersonal experts. The successful use of the Team Manifesto and the Process Check depends on our other recommendations. Structured, well-defined practices, rules of engagement, and moderators interrupt bias and ensure productive collaboration. Well-defined team onboarding ensures that everyone knows what to do and is set up to be successful. Well-designed processes ensure that women are treated equally, and that a diverse team works well. But we must also have a way to address interpersonal friction outside of meetings. If we do not address valuing, jerk, and professional behavior, the team will be faced with interpersonal chaos. We will not get the best from team members. The Valuing and Jerk Workshop and the Team Manifesto make sure that teams take responsibility for how they behave and treat each other. Reflection in a Process Check is only effective because we have already redesigned our practices and explicated acceptable behaviors.

> *Create resilient, high performing, diverse teams by designing how they work together*

The Team Manifesto and Process Check also enable the team to own and manage their culture. They are key to creating a local business culture that can supersede their interpersonal differences. Even in a company with a potentially hostile culture, teams can create an island of respect and safety. When this happens, the impact is enormous. The team can become valuing, dynamic, and up to something big producing innovation. Resilient teams, dynamic diverse teams, don't just help women—they ensure the core business goals.

Part II Conclusion

In Part II, we have introduced tested interventions that help diverse teams improve the way they work. Our interventions structure core practices so they work better for women and all team members. They help remove interpersonal chaos and ensure that all members have a voice. They foster a valuing culture that can use feedback well. Our three-pronged attack to retain women in tech is informed by the needs expressed in The @Work Experience Framework. But our solutions are designed to work for the whole team. By focusing on redesigning practices, interpersonal dynamics, and team values; by becoming a reflective self-correcting organization, companies can start to create the type of work experience that women desire. These practices are at the heart of retaining women in tech.

Next, in the Conclusion, we summarize the principles behind our approach to redesigning practices to ensure that women—and everyone—thrives in tech companies.

Principles of Process Intervention for Retaining Women in Tech

We started the Women in Tech Retention Project to understand why women are leaving our field twice as often as men. Through a combination of field research and surveys, we identified the factors in The @Work Experience Framework, which point organizations to the experiences they need to create to retain women. Our research was focused on women across multiple job and company types. Both our qualitative and worldwide quantitative research included data from people of color, people who identify as LGBTQ+, and men. We think the findings and suggestions presented in this book are a good place to start thinking about what your need for your organization or team.

In Part I, we shared the key factors that help women thrive in tech. The @Work Experience Framework gives you a "pair of glasses" from which to investigate what is going on in your workplace. It helps you focus on the attributes and experiences in the daily work life of people in your organization and whether it works for all the members of your diverse teams. Because the Framework represents an overarching view of women's experience at work, it can help you see the big picture of what you need to consider for retention. It also identifies specific factors that make up this experience and gives you specific areas to investigate and improve in your organization. As a reminder:

Dynamic, Valuing Team. Women thrive in a dynamic, work-focused team and/or partnership where they can lead, follow, feel valued, and talk about life outside of work within their team.

Stimulating Work. Women love working on challenging technical problems, products, or research questions important to the company, the industry, or the world. They switch jobs when bored.

Push & Support. Women may not feel qualified for the next challenge. But when pushed by trusted managers, colleagues, or family they take it on and succeed—provided they have support to strategize with others, ask questions, and falter.

Local Role Models. Women need coaching relationships in their company to help them succeed. Work buddies with more experience help them navigate their careers. Senior co-workers and managers also reveal the experience of daily life after promotion. If those lives look undesirable, women may not seek promotion.

Nonjudgmental Flexibility. Women with children thrive if the team and managers flex to everyone's life commitments. These women too often feel judged for meeting home commitments. Being given flexibility by the team and managers shows them **that they are valued.**

Personal Power. Women can have self-doubt about their skills, readiness, and value. Self-esteem increases with positive feedback, helpful critique, clear expectations, and good coaching.

The @Work Experience Framework outlines the experiences that matter for retention. But addressing only one of these factors will not be enough; all six factors work together to form the working experiences for women. Since we have started our work on retaining women in tech, the world of work has changed. Because of the pandemic, many tech workers have been working remotely alone in their homes. Throughout the book, we have shared findings and solutions related to remote workers. Let's look at remote working from the point of view of creating the experiences women need to thrive.

REMOTE WORKING AND THE @WORK EXPERIENCE FRAMEWORK

Remote working during the pandemic goes far beyond the global remote work that occurred in prior years, and about which many researchers have studied. The unique nature of the pandemic posed challenges to what we knew before. The whole tech world became a living lab with people working alone and exclusively interacting through remote video, collaborative whiteboards, and other collaboration tools. Organizations learned that the work can indeed get done if people work remotely. Some are choosing to move to an all-remote work culture. Many others are considering some kind of hybrid model. As of this writing, we have not returned to the workplace en masse. There will be many organizational experiments over the next few years. So, it makes sense to ask first, if working remotely is good for women and second, if our recommendations and perspectives still hold in a remote work context.

To understand the impact of remote work on women today, we launched The Remote Work Project doing in-depth interviews with women in tech in a variety of job roles. We also included open-ended questions on remote work in one of our surveys. Findings from both sources dovetail. It is still too early to understand what will ultimately happen for women and work practices as remote working evolves. But our findings are suggestive. The "glasses" of the Framework help us anticipate and be on the watch for issues. Our current findings from The Remote Work Project and other research[1, 4, 5] revealed issues that will impact women. We summarize the impact of remote work below.

Creating connections. People have an increased difficulty getting to know each other both socially and as work partners. The lack of easy connection impacts team cohesion, building new relationships with work partners and managers, and enjoying the natural chitchat that allows for breaks in a co-located workplace. Remote working makes it hard for new hires to make relationships and develop a sense of belonging. The difficulty of creating and maintaining connections for all workers has a direct impact on forming Dynamic Valuing Teams.

Getting help and coaching. People, and especially new hires, feel that it is more difficult to get one-on-one help from managers and co-workers. Scheduling one-on-one meetings feels more like an intrusion when remote. New hires don't want to impose given how busy everyone is. Asking for time when it involves setting up a meeting is harder. In general, drop-in questions and help are not easy when remote. This finding tells us that working with Local Role Models and getting a Push and Support may be harder when remote.

Stimulating work. The above two findings raise the worry of whether women working remotely will get the stimulating work they want. If women receive less coaching, will they advance their skills? If they do not feel supported because it is harder to make relationships all around, will they be less likely to put themselves forward for challenging work? In a remote context, will women's contributions be less apparent to managers and coworkers? With less daily interaction will our bias against women's skills in tech disadvantage women even more?

Communicating informally. Informal communication outside of meetings suffers when everyone is remote. Side conversations, which used to happen before and after meetings and in the hallway, have nearly evaporated. Also gone are the informal sharing of work and helping each other when stuck that were common practices when co-located. People are surprised by how much collaboration depended on dropping in. They did not realize that informal interactions drove forming shared understandings of what to do or decide. The people we interviewed reported both errors and time wasted by formal requests through systems. Setting up meetings to chat does not substitute for dropping in and seems like an imposition. More importantly, this informal collaboration is invisible; people don't think about how to replace something when they don't know it is happening. It is that much harder to talk things over, to be sure you know what to do by checking in with others, and to create the overall connections that happen when people combine social and work talk without thinking about it. The work, the relationships, and the sense of collaborating on something that really matters all suffer. Remote working makes it is harder for teams to collaborate dynamically, which impacts being a Dynamic, Valuing Team.

Structuring practices. We find that managers and teams have introduced more structure to their remote group working meetings. When remote, less structure and more freeform interactions in group working meetings have led to the experience of time being wasted, confusion in managing participation, and the work not getting done. Senior people have naturally gravitated to increasingly explicit practices and using remote collaborative tools to make the purpose and interactions in the meetings clear. Indeed, research finds that more structure in remote collaboration heightens the collective intelligence of the team.[8] And more structure is good for women because the implicit becomes explicit, interactions are potentially better managed, and practices are clear. As we return to more face-to-face interactions, we hope that managers and teams will bring this increased structure with them.

Supporting life flexibility. People have emphasized that during the pandemic their manager and teams have been more willing to flex when home needs appear. We are quite happy that teams do seem to be more understanding of home/work commitments when remote. The pandemic highlighted how the crush of home and family demands can overwhelm—and that working at home can reduce the burden because home chores and childcare can be interleaved with working. But as numerous others have pointed out, during the pandemic the burden of children and the home has fallen within gender norms; too many women are struggling to do the work and deal with home life.[3] Until all schools, daycare, and elder care are safe and back in place, we will not know if Nonjudgmental Flexibility will continue and how the balance of responsibility will shift as we move to more in-person work. This will have to be watched.

Increasing self-confidence. Given our overall findings about remote work, Personal Power and self-confidence will be affected. Without connection, coaching, help, and informal information, the opportunity for misunderstanding and accomplishments being overlooked is great. But we are happy to report that some managers, teams, and companies are deliberately trying to be more valuing, one even sending weekly notes to encourage employees to send valuing messages. Self-confidence is also related to feeling psychologically safe. Interestingly, some women report feeling safer to express themselves behind the screen, more willing to participate or take a risk. And they report using side chats in a meeting to check in with each other and send notes of support if interpersonal issues get out of control. Of course, this works better when the camera is on, and everyone can see each other's nonverbal responses. Overall, self-confidence is clearly impacted in remote settings.

Keeping cameras on. We know from years of research on remote meetings that video on sends more cues to guide interactions than video off. But it's complicated. Most people are more comfortable with their own camera off. At the same time, people often would like to see the others in their meetings. Negotiating this contradiction is one of the challenges of remote meetings. And there is no right or wrong. There are, however, considerations regarding women that play a role here. We have some indication that keeping the video on in a remote setting may help women feel more able to speak up. At the same time, being on camera puts pressure on women to look good that men do not have in the same way.[9] But more of our findings surround what happens with cameras off. Camera off choices can be policies designed to respect individual choices. In meetings, some people are on video while others are not, a personal choice. But we also see people following the lead of their managers. We find that if people are new and don't know co-workers it is much harder to get to know them with the camera off.[6] And some of the women we talked to report that getting negative feedback remotely with the camera off can come across harsher. From an interpersonal dynamics perspective, the camera on vs. off clearly changes the experience of working meetings. As we become more permanently remote, we will have to examine if there is a gender effect and the real impact on team cohesion and connection.

These are not all the findings from our remote project, but they are the high-level themes. Many people like the flexibility of being able to work from home at least some of the time. They like the control over their time, and they love not having a commute. But there is no question that being remote changes the overall work experience and creates new challenges. Some research even suggests that the standard speech compression methods that are used in video-conference tools cause women's voices to sound less competent.[7] This is bad news considering the already existing bias against women's competence. We will learn more as more experiments in remote working are conducted by our industry.

Stepping back, our research on remote experiences only makes attention to the issues raised in The @Work Experience Framework more important. A good team onboarding process is even more important when new hires and teams are remote. If all new hires, including women, do not get a good start, ensuring their success when remote will be even harder. Since more can be misperceived when remote, increasing structure when giving feedback in the processes used in working meetings is essential. Remote work needs even more explicit practices, rules of engagement, procedures to manage participation, and good facilitation. Because interpersonal connection is more difficult, remote work requires even more attention to interpersonal dynamics. Increasing valuing behaviors and managing devaluing behaviors is a must in remote interactions. As for jerk behavior, we do not find more or less of it when remote. It seems that people who behave non-professionally when face-to-face also engage in these behaviors when remote, but it is more apparent to others, who may intervene. In other words, with remote work the daily work life changes, but the challenges for retaining women do not. Whether we are working face-to-face or remotely women need the same things to thrive. To make that happen, we need to deliberately design interventions to help.

PRINCIPLES OF INTERVENTION DESIGN

In Part II we shared interventions that holistically address areas for change in women's experiences at work. The interventions redesign key processes used by technology teams, interpersonal interactions, and building and maintaining a valuing team culture. These interventions were identified as important by our research. They are also central to working in technology groups. The interventions are not special for women; they seek to improve the practice for everyone on the team with an eye toward issues of women and diverse teams. We have found that addressing how diverse teams work together can effectively interrupt and manage stereotyping and bias, foster

team cohesion, and get the work done with greater quality. We do not try to change the behaviors and attitudes of individual people directly through bias training, awareness workshops, or other techniques. Instead, our interventions impact individual behavior by redesigning how teams work and interact. As a reminder:

> **Team Onboarding Checklist.** 49% of women in our @Work Experience survey do
> not know what to do to be successful. By explicating expectations for new hires, we
> help managers know how to help new hires build critical connections and gain the
> skill and knowledge they need to succeed. By putting these activities into a checklist,
> we make the practice concrete and so easy to execute against. By stating what new
> hires need, we encourage managers and teams to explicate implicit elements of success
> like team culture, criteria for success, expectations of a job role, and more. By stating
> exactly what kind of relationships new hires need, we nudge managers to make these
> connections happen. By explicating what is needed to create success for a new hire, we
> ensure that new hires, including women, will get connected and succeed.

The Critique Meeting. 80% of women lose self-confidence when criticized. The Critique Meeting puts structure into how feedback meetings are run, ensuring that feedback is given in a way that can be received by those being reviewed. We introduce rules of engagement, criteria for goodness, and clarity on how to participate so that feedback is professional and not personal. By using a round-robin structure for gathering feedback, we ensure that all voices are heard and that no voice dominates. And by making it clear that the purpose of the meeting is to give feedback, not fix problems, we ensure that the meeting achieves its goals and does not devolve into chaos. All of these techniques directly address issues that early-career professionals and diverse populations may have when receiving critique. All of these techniques also work for everyone and ensure a high-quality work product.

Scrum. Scrum is the most popular form of Agile software development method used in tech companies. It is already highly structured with clear working meeting practices. But this does not mean that it necessarily works well for women and diverse team members. Our analysis of Scrum reveals ways to tune its practices by introducing additional structure to ensure that discussions are better managed, and all voices are heard no matter their gender or level of experience or influence. Small changes in how conversations are conducted or how tasks are "chosen," tune the practice, and smooth both the work process and collaboration.

Valuing and Jerk Behaviors. These are the most desired and loathed behaviors that impact women's experience of value and belonging. To address them, we started with research to define exactly which behaviors are desired and undesired. We then named behavior clusters with fun, non-threatening characters, making the behaviors explicit. Characterizing the behaviors in fun posters allows people and teams to discuss what they need to do to increase valuing behaviors. For jerk behaviors, individuals can self-recognize and make choices to change by finding a substitute behavior. Making the behaviors explicit helps teams to tune their code of conduct and allows for facilitation in working meetings. Self-recognition allows individuals to define personal growth goals and get help to achieve them. Awareness of behavior plus explicit processes to hold ourselves accountable can change behavior.

Team Manifesto and Process Checks. The Team Manifesto is a practice to help teams explicitly define their values and expectations. The process to create a manifesto includes exercises to raise up issues that matter for women and diverse teams to prime reflection on team culture. The process to create the manifesto is also well-defined, ensuring that the team shares and commits to their definition of expected behaviors and values. The manifesto publicly declares the team's values, code of conduct, and work expectations. But listing expectations alone is not enough to ensure accountability. Frequent Process Checks build in reflection, another explicit process, to identify how the team is doing and what they need to do to tune their practices and interactions.

These interventions are practical changes that managers and teams can use and tune to improve how diverse teams work. They are also a model of how to think about redesigning practices to improve the daily work life of diverse teams. Throughout Part II we have referenced our principles of process redesign to improve the work experience of women in tech. Below, we bring these principles together to help guide your own intervention design.

DO USER-CENTERED PROCESS REDESIGN

Redesigning processes is not much different than designing products that work for users. For every process redesign, we begin with interviews and observations of how existing processes are working. We start with those processes that are core to the overall work experience of diverse teams, those related to the necessary experiences of The @Work Experience Framework, and those where organizations receive numerous complaints. Once a process is selected, start by finding out what is happening. Karen uses Contextual Design techniques. Nicola uses a Living Lab approach. But whichever user-centered design technique you use the components of process redesign are the same:

1. Do field interviews with multiple people on a variety of teams.

2. Organize the data to bubble up themes and issues.

3. (Optional) Use this data to create a survey of a larger population or with more groups in your organization. This helps determine the overall population experience and where the real pain points are.

4. Run ideation sessions where participants "walk" the data and generate changes to the processes to improve them for everyone with an eye toward the needs of diverse team members.

5. Include a few people who know how to think about process design in the ideation session along with a selection of managers and representatives from diverse populations, including white men.

6. Mockup tools or procedures to address the issues.

7. Iterate the tool or procedure with real teams and individuals.

8. Finalize the new process and roll it out.

9. Keep observing and reflecting to tune the process in use.

Outside consultants can help you do the work but don't give all responsibility for understanding and intervention design to them. Stay involved throughout as members of the change process. This is the best way to start to build buy-in. This kind of participatory process design is always better than introducing a new practice without input. Start with our ideas as templates; iterating on something that is already proven is always better than starting from scratch. A user-centered approach is always the best starting place for process redesign. It helps you understand the experiences of the people in your organization and be honest about what is happening for diverse people who work there.

MAKE THE IMPLICIT EXPLICIT

At the core of all of our recommendations is our central principle of redesign: make implicit values, expectations, and practices explicit and deliberately chosen. All of our interventions make what is implicit, explicit; what is unarticulated, articulated; what is driven by bias, open to examination. When a practice, value, or expected behavior is known and chosen, facilitation and Process Checks are possible. But when interactions are organic, they are more likely to encourage chaos, blame, and devaluing. Being explicit makes us more accountable and allows for practice redesign when it is not working well. Explicit practices, expectations, and values allow everyone to succeed; all know the rules of engagement. In other words, explicit structure encourages professionalism and disrupts our tendency toward bias—always better for women and diverse teams. We can't make everything explicit. But we can get much closer than we are today.

Any organization can make its processes and practices more explicit and therefore less prone to bias and confusion about what to do to be successful. We introduced The Analysis Matrix to help organizations identify the implicit and explicit aspects of their practices. The Analysis Matrix is not just a tool to think about and redesign Scrum practices—it is a tool that can be used for any process to find implicit practices that may be opening the door to bias and stereotyping. We encourage you to use it and the stimulus questions to examine any process that undercuts or silences women and other underrepresented populations.

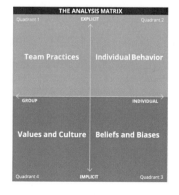

Finding the implicit can help you determine where you want to deliberately redesign your practices and culture. As we have said, the more we explicate, redesign, and monitor our practices the more we will uncover implicit practices, expected behaviors, values, and attitudes that unintentionally hinder women and diverse team members. Then diverse teams will be more likely to deliver peak performance and innovative ideas—and we will retain more women in tech.

STRUCTURE WORKING MEETINGS

The chief characteristic of a maker culture is collaboration between individuals and in working meetings. Here the principles guiding explicit practices also may be applied. Lack of structure in meetings has a negative impact on the team and the organization's bottom line. For example, Doodle's State of Meetings Report 2019 estimated that poorly organized meetings in the United States, Germany, Switzerland, and Austria cost $564 billion in 2019.[2]

We use The Critique Meeting as an example of one kind of working meeting that can be made explicit. We examine Scrum because its success is partially due to well-designed working meetings. We then improve Scrum by tuning how some aspects of the meetings work. Well-designed working meetings make practices and expectations explicit both for doing the work and interacting.

How meetings are run is important to women and retention because of their profound impact on communications of value, self-esteem, feelings of belonging, and whether work and skill are acknowledged. Well-run working meetings help deal with interpersonal chaos, conflict, and inadvertent hurt. They also can manage bad and unprofessional behavior. Well-run recurring working meetings also set the stage for how people treat each other in general. As we said about The Critique Meeting, feedback can be difficult to take. Establishing expectations of *how* to provide feedback by using well-designed meetings can also create a habit of how to provide feedback in one-on-one interactions. Well-designed working meetings can nudge team members toward more professional and productive interactions.

As a reminder, well-designed working meetings include:

- Meeting owners who communicate clear goals and expectations for the meeting
- Well-defined processes including clear goals, roles, processes, and expected outcomes and structured participation techniques
- An artifact as a focal point for all to "see" what is being created or agreed to
- A small enough number of people so everyone has airtime
- A moderator for groups of > 3 people
- Articulated rules of engagement relevant to that specific type of working meeting
- Shared principles of goodness relevant to decisions being made

Use these dimensions of well-run working meetings to examine key meetings in your organization. And remember the principle of separating conversations. You can design working meetings around a single coherent conversation as occurs in Scrum. But within a meeting you can listen for and facilitate separation of different conversations, which so often leads to conflict. Even with good practices, the role of a knowledgeable moderator goes a long way to ensuring both that working meetings are productive and that everyone is heard.

NAMING AND CLAIMING—NOT BLAMING

The hardest change to make is in how people treat each other—interpersonal dynamics. Naming a behavior is the way we make behaviors explicit and recognizable. We shared the power of naming in Chapter 10 with Valuing and Jerk Behaviors and how fun can defuse discussion of behaviors. Change starts with naming. If we can't say exactly which behaviors are problematic, we can't design interventions to increase or decrease them. But then how do we address change? We are guided by another principle of behavior change: self-recognition and personal growth goals. Blaming others simply stimulates anger, resentment, denial, and defensiveness. But naming raises up behaviors so that they can be seen. If we let them, people can self-recognize and commit to their own change through personal growth goals. Teams can recognize what is in their way and create values and codes of conduct, or manifestos that allow them to reflect and self-modulate.

This is not to say that managers, codes of conduct, and review processes cannot more directly address behavior change and expectations. Nor does it mean that when something untoward happens that we should not raise it to the person themselves or a manager. But if teams work together to self-identify personal growth goals and also deliberately create a code of conduct and process to monitor success, they may ensure these noxious behaviors do not occur in the first place. No one is perfect, apology is powerful.

This is our preferred way to build accountability and professional behavior into the daily interactions. *Naming and claiming—not blaming is a core principle of intervention.*

DESIGN FOR EASY ADOPTION

Redesigning practices only works if managers and teams use the new practices. Including people in the redesign and iteration process is always a good approach to building buy-in. But beyond involvement in the redesign process, we guide our interventions by the principle of easy adoption. Our goal is to make it as painless as possible for people to try new approaches and iterate them to work for their teams. When we tune a practice, we can "sneak" it into how people work. If we make adoption easy, we might create bottoms-up change without needing organizational blessing. Being sneaky in creating change means attention to these principles below:

1. Design changes at the level of the manager and the team. If the practice does not need to be adopted by the organization as a whole, it has a greater chance of slipping in and starting a grassroots movement.

2. Keep it simple. If you are making a tool like a checklist or rules of engagement, make sure it can be used immediately, requires no training, and works in both in-person and remote contexts.

3. Fit into and tune existing organizational processes. Start at the level of the team. Don't boil the ocean and try to redesign the entire culture or every process in the organization.

4. Be sure that any team changes can be integrated into existing corporate tools and iterated to work for the organization.

5. Make sure that your intervention addresses issues raised by women in tech but doesn't target them. Creating specific processes just for women signals them out as "special," which can create resentment and play into already negative stereotypes. The intervention must work for everybody.

As you redesign your practices, keep simple adoption in mind. Stay low-key. The Team Onboarding Checklist works for everyone, as do all our interventions. Focus on improving practices with an eye toward gender issues. But we can only retain women in tech if all members of the diverse team thrive and succeed.

The principles we have shared here can help guide the changes to your practices and culture that can help retain women in tech. Our principles depend on a deep understanding of women's experiences in your organization and then deliberate and explicit design of small changes that fit easily into daily work life. We can raise up expected interpersonal behavior explicitly, but then we must build new expectations into our practices, reflections, and values. Through continuous reflection, we can improve how we work and interact, which helps build a valuing resilient culture. Everyone has some implicit bias whether we want to or not. We might not be able to change how we are wired, but if we are well-meaning—and most people are—we can manage those impulses by design.

Women face real barriers, bias, and a tightrope walk trying to deal with expectations installed by gender stereotypes and what's expected within the tech culture. Throughout this book, we have shared the realities of what women in tech face, and what they need to thrive. Women leave jobs and the tech industry when they do not find it welcoming and the work life livable. But organizations, managers, and teams can deliberately design their work experiences so women and underrepresented people feel valued, connected, and achieve success. If we, as an industry, do that, diverse teams will be cohesive and dynamic. We will be able to call upon all the diverse skills and perspectives on the team to make great products. And we will retain women in tech.

References

INTRODUCTION

[1]   Joan Acker. (2006). Inequality Regimes: Gender, Class, and Race in Organizations. *Gender & Society*, 20(4), 441–464. doi:10.1177/0891243206289499. 8

[2]   Michael Ahmadi, Anne Weibert, Rebecca Eilert, Volker Wulf, and Nicola Marsden. (2020). Feminist Living Labs as Research Infrastructures for HCI: The Case of a Video Game Company. *Proceedings of the 2020 CHI Conference on Human Factors in Computing Systems*, 1–15. doi:10.1145/3313831.3376716. 20

[3]   Michael Ahmadi, Anne Weibert, Corinna Ogonowski, Konstantin Aal, Kristian Gäckle, Nicola Marsden, and Volker Wulf. (2018). Challenges and Lessons Learned by Applying Living Labs in Gender and IT Contexts. *4th Gender & IT Conference (GenderIT'18)*, 239–249. doi:10.1145/3196839.3196878. 20

[4]   Michael Ahmadi, Anne Weibert, Victoria Wenzelmann, Konstantin Aal, Kristian Gäckle, Volker Wulf, and Nicola Marsden. (2019). Designing for Openness in Making: Lessons Learned from a Digital Project Week. *9th International Conference on Communities & Technologies - Transforming Communities (C&T '19)*, 160–171. doi:10.1145/3328320.3328376. 10

[5]   Teresa Almeida, Marie-Louise Juul Søndergaard, Sarah Homewood, Kellie Morrissey, and Madeline Balaam. (2018). Woman-Centred Design. In Sharon Prendeville, Abigail Durrant, Nora O'Murchú and Keelin Leahy (Eds.), *DRS 2018: Book of DRS 2018 Conversations UK*. doi:10.21606/drs.2018.795. 5

[6]   Hala Annabi and Sarah Lebovitz. (2018). Improving the Retention of Women in the IT Workforce: An Investigation of Gender Diversity Interventions in the USA. *Information Systems Journal*, 28(6), 1049–1081. doi:10.1111/isj.12182. 12

[7]   Catherine Ashcraft, Brad McLain, and Elizabeth Eger. (2016). NCWIT - Women in Tech: The Facts. *NCWIT*. Retrieved from https://www.ncwit.org/sites/default/files/resources/ncwit_women-in-it_2016-full-report_final-web06012016.pdf. 2, 6

[8]   Mahzarin R Banaji and Curtis D Hardin. (1996). Automatic Stereotyping. *Psychological Science*, 7(3), 136–141. doi:10.1111/j.1467-9280.1996.tb00346.x. 12

[9]   Mahzarin R. Banaji, R. Bhaskar, and Michael Brownstein. (2015). When Bias Is Implicit, How Might We Think About Repairing Harm? *Current Opinion in Psychology*, 6, 183–188. doi:10.1016/j.copsyc.2015.08.017. 12

[10]   Corinna Bath. (2014). Searching for Methodology. Feminist Technology Design in Computer Science. In Waltraud Ernst and Ilona Horwath (Eds.), *Gender in Science and Technology* (pp. 57–78). Bielefeld: transcript. doi:10.14361/transcript.9783839424346.57. 4

[11]   Marianne Bertrand and Sendhil Mullainathan. (2004). Are Emily and Greg More Employable Than Lakisha and Jamal? A Field Experiment on Labor Market Discrimination. *American Economic Review*, 94(4), 991–1013. doi:10.1257/0002828042002561. 14

[12]   Iris Bohnet. (2016). *What Works: Gender Equality by Design*. Cambridge, Massachusetts: Harvard University Press. doi:10.4159/9780674545991. 14, 15

[13]   Clint A. Bowers, James A. Pharmer, and Eduardo Salas. (2000). When Member Homogeneity Is Needed in Work Teams: A Meta-Analysis. *Small Group Research*, 31(3), 305–327. doi:10.1177/104649640003100303. 4

[14]   Emilio J. Castilla and Stephen Benard. (2010). The Paradox of Meritocracy in Organizations. *Administrative Science Quarterly*, 55(4), 543–676. doi:10.2189/asqu.2010.55.4.543. 7

[15]   Catalyst. (2013). Why Diversity Matters. Retrieved from http://www.catalyst.org/system/files/why_diversity_matters_catalyst_0.pdf. 4

[16]   Katy Cook. (2020). *The Psychology of Silicon Valley*. Cham: Palgrave Macmillan. doi:10.1007/978-3-030-27364-4. 7

[17]   David A. Cotter, Joan M. Hermsen, and Reeve Vanneman. (2004). *Gender Inequality at Work*: Russell Sage Foundation New York. 2

[18]   Susan Davis-Ali. (2017). Advancing Women Technologists into Positions of Leadership. *Anita Borg Institute*. Retrieved from https://anitab.org/wp-content/uploads/2020/08/advancing-women-technologists-leaders.pdf. 6

[19]   Megan Rose Dickey. (2017). Dispatches on Diversity: Uber, Sexual Harassment and Venture Capital. Retrieved from https://techcrunch.com/2017/09/23/dispatches-on-diversity-uber-sexual-harassment-and-venture-capital/. 6

[20]   Frank Dobbin and Alexandra Kalev. (2018). Why Doesn't Diversity Training Work? The Challenge for Industry and Academia. *Anthropology Now*, 10(2), 48–55. doi:10.1080/19428200.2018.1493182. 12

[21] Arielle Duhaime-Ross. (2014). Apple Promised an Expansive Health App, So Why Can't I Track Menstruation? Retrieved from https://www.theverge.com/2014/9/25/6844021/apple-promised-an-expansive-health-app-so-why-cant-i-track. 5

[22] Charles Duhigg. (26. Feb. 2016). What Google Learned from Its Quest to Build the Perfect Team. *The New York Times Magazine*. Retrieved from https://www.nytimes.com/2016/02/28/magazine/what-google-learned-from-its-quest-to-build-the-perfect-team.html. 16

[23] Naomi Ellemers and Manuela Barreto. (2009). Maintaining the Illusion of Meritocracy: How Men and Women Interactively Sustain Gender Inequality at Work. In S. Demoulin, J.-P. Leyens and J. F. Dovidio (Eds.), *Intergroup Misunderstandings: Impact of Divergent Social Realities* (pp. 191–212): Psychology Press. 7

[24] Eurostat. (2021). Employed ICT Specialists by Sex. Retrieved from https://appsso.eurostat.ec.europa.eu/nui/show.do?dataset=isoc_sks_itsps&. 2

[25] Carol Farnsworth and Karen Holtzblatt. (2016). Diversity in High Tech: Retaining Employees Once They're in the Door. *Proceedings of the 2016 CHI Conference Extended Abstracts on Human Factors in Computing Systems*, 1077-1080. doi:10.1145/2851581.2886429. 9, 18

[26] Seth Fiegerman. (2017). Labor Department Goes after Big Tech for Discrimination. Retrieved from http://money.cnn.com/2017/04/10/technology/labor-department-tech/index.html. 6

[27] Cary Funk and Kim Parker. (2018). Women and Men in STEM Often at Odds over Workplace Equity. *Pew Research Center*. Retrieved from http://assets.pewresearch.org/wp-content/uploads/sites/3/2018/01/09142305/PS_2018.01.09_STEM_FINAL.pdf. 3, 6

[28] Denise L. Gammal and Caroline Simard. (2013). Women Technologists Count - *Anita Borg Institute Solutions Series*. Retrieved from https://diversity.hrtechgroup.com/sites/default/files/2020-06/Women_Technologists_Count %281%29.pdf. 2

[29] Danielle Gaucher, Justin Friesen, and Aaron C. Kay. (2011). Evidence That Gendered Wording in Job Advertisements Exists and Sustains Gender Inequality. *Journal of Personality and Social Psychology*, 101(1), 109–128. doi:10.1037/a0022530. 2, 16

[30] Connie J. G. Gersick. (1988). Time and Transition in Work Teams: Toward a New Model of Group Development. *Academy of Management Journal*, 31(1), 9–41. doi:10.5465/256496. 16

[31]   Jennifer L. Glass, Sharon Sassler, Yael Levitte, and Katherine M. Michelmore. (2013). What's So Special About STEM? A Comparison of Women's Retention in STEM and Professional Occupations. *Social Forces*, 92(2), 723–756. doi:10.1093/sf/sot092. 3

[32]   Claudia Goldin and Cecilia Rouse. (2000). Orchestrating Impartiality: The Impact of" Blind" Auditions on Female Musicians. *American Economic Review*, 90(4), 715–741. doi:10.1257/aer.90.4.715. 14

[33]   Marc Goulden, Mary Ann Mason, and Karie Frasch. (2011). Keeping Women in the Science Pipeline. *The ANNALS of the American Academy of Political and Social Science*, 638(1), 141–162. doi:10.1177/0002716211416925. 2

[34]   Anthony G. Greenwald and Mahzarin R. Banaji. (2017). The Implicit Revolution: Reconceiving the Relation between Conscious and Unconscious. *American Psychologist*, 72(9), 861–871. doi:10.1037/amp0000238. 12

[35]   Madeline E. Heilman, Francesca Manzi, and Susanne Braun. (2015). Presumed Incompetent: Perceived Lack of Fit and Gender Bias in Recruitment and Selection. In Adelina M. Broadbridge and Sandra L. Fielden (Eds.), *Handbook of Gendered Careers in Management: Getting in, Getting on, Getting Out* (pp. 90–104). Cheltenham: Edward Elgar. doi:10.4337/9781782547709.00014. 6

[36]   Cedric Herring. (2009). Does Diversity Pay? Race, Gender, and the Business Case for Diversity. *American Sociological Review*, 74(2), 208–224. doi:10.1177/000312240907400203. 4

[37]   Sylvia Ann Hewlett, Carolyn Buck Luce, Lisa J. Servon, Laura Sherbin, Peggy Shiller, Eytan Sosnovich, and Karen Sumberg. (2008). The Athena Factor: Reversing the Brain Drain in Science, Engineering and Technology. *Center for Work-Life Policy*. Retrieved from https://store.hbr.org/product/the-athena-factor-reversing-the-brain-drain-in-science-engineering-and-technology/10094. 2

[38]   Karen Holtzblatt, Aruna Balakrishnan, Troy Effner, Emily Rhodes, and Tina Tuan. (2016). Beyond the Pipeline: Addressing Diversity in High Tech. *Proceedings of the 2016 CHI Conference Extended Abstracts on Human Factors in Computing Systems*, 1063–1068. doi:10.1145/2851581.2886422. 9, 18

[39]   Karen Holtzblatt and Hugh Beyer. (2015). *Contextual Design Evolved*. San Rafael, CA: Morgan & Claypool Publishers. 8

[40]   Karen Holtzblatt and Nicola Marsden. (2018). Retaining Women in Technology - Uncovering and Measuring Key Dimensions of Daily Work Experiences. *IEEE Inter-*

national Conference on Engineering, Technology and Innovation (ICE/ITMC), 148-155. doi:10.1109/ICE.2018.8436351. 9, 10, 18

[41] Honeypot. (2018). 2018 Women in Tech Index. Retrieved from https://www.honeypot.io/women-in-tech-2018/. 2

[42] Jenny M. Hoobler, Courtney R. Masterson, Stella M. Nkomo, and Eric J. Michel. (2018). The Business Case for Women Leaders: Meta-Analysis, Research Critique, and Path Forward. *Journal of Management*, 44(6), 2473–2499. doi:10.1177/0149206316628643. 4

[43] Vivian Hunt, Sara Prince, Sundiatu Dixon-Fyle, and Lareina Yee. (2018). Delivering through Diversity. *McKinsey*. Retrieved from https://www.mckinsey.com/business-functions/organization/our-insights/delivering-through-diversity. 3, 4

[44] Iclaves. (2013). Study on Women Active in the ICT Sector - a Study Prepared for the European Commission. *Publications Office of the European Union*. Retrieved from http://bcwt.bg/wp-content/uploads/documents/womenActiveInICT.pdf. 2, 3

[45] Level Playing Field Institute. (2011). *The Tilted Playing Field: Hidden Bias in Information Technology Workplaces*. San Francisco, CA: Kapor Center. 7

[46] Charlotte Jee. (2018). Sexism in the Tech Sector: The Biggest Stories. Retrieved from https://www.techworld.com/careers/sexism-in-tech-sector-biggest-stories-3670259/. 6

[47] June Park John and Martin Carnoy. (2019). The Case of Computer Science Education, Employment, Gender, and Race/Ethnicity in Silicon Valley, 1980–2015. *Journal of Education and Work*, 32(5), 421–435. doi:10.1080/13639080.2019.1679728. 2

[48] Stacey Jones and Jaclyn Trop. (2015). See How the Big Tech Companies Compare on Employee Diversity. Retrieved from http://fortune.com/2015/07/30/tech-companies-diveristy/. 2

[49] Damien Joseph, Soon Ang, and Sandra A. Slaughter. (2015). Turnover or Turnaway? Competing Risks Analysis of Male and Female IT Professionals' Job Mobility and Relative Pay Gap. *Information Systems Research*, 26(1), 145–164. doi:10.1287/isre.2014.0558 3.

[50] Rosabeth Moss Kanter. (2008 (1977)). *Men and Women of the Corporation*: New Edition: Basic Books. 4

[51] Andre Kaspura. (2019). The Engineering Profession: A Statistical Overview - Engineers Australia, 14th Edition. Retrieved from https://www.engineersaustralia.org.au/Government-And-Policy/Statistics. 2

[52] Julian Kolev, Yuly Fuentes-Medel, and Fiona Murray. (2019). Is Blinded Review Enough? How Gendered Outcomes Arise Even under Anonymous Evaluation. *National Bureau*

of Economic Research. Retrieved from https://www.nber.org/system/files/working_papers/w25759/w25759.pdf. doi:10.3386/w25759. 14

[53] Jennifer LaCosse, Denise Sekaquaptewa, and Jill Bennett. (2016). STEM Stereotypic Attribution Bias among Women in an Unwelcoming Science Setting. *Psychology of Women Quarterly*, 40(3), 378–397. doi:10.1177/0361684316630965. 6

[54] Liana Christin Landivar. (2013). Disparities in STEM Employment by Sex, Race, and Hispanic Origin. *Education Review*, 29(6), 911–922. 2

[55] Julia Malacoff. (2017). 12 Cool Companies with Amazing Maternity Leave Policies. Retrieved from https://www.glassdoor.com/blog/12-cool-companies-with-amazing-maternity-leave-policies/. 7

[56] Katherine Cumings Mansfield, Anjalé D. Welton, and Margaret Grogan. (2014). "Truth or Consequences": A Feminist Critical Policy Analysis of the STEM Crisis. *International Journal of Qualitative Studies in Education*, 27(9), 1155–1182. doi:10.1080/09518398.2014.916006. 1

[57] Nicola Marsden. (2019). *Wer entscheidet, wie die Welt wird? Gender Bias in der IT*. Fernuniversität Hagen: Keynote Talk at GenderDay 2019. 5

[58] Nicola Marsden and Karen Holtzblatt. (2018). How Do HCI Professionals Perceive Their Work Experience? Insights from the Comparison with Other Job Roles in IT. *Extended Abstracts of the 2018 CHI Conference on Human Factors in Computing Systems*, LBW522: 521-526. doi:10.1145/3170427.3188501. 9

[59] Kenneth Matos, Ellen Galinsky, and James T. Bond. (2017). National Study of Employers. *Society for Human Resource Management*. Retrieved from http://whenworkworks.org/downloads/2016-National-Study-of-Employers.pdf. 7

[60] Kimberly McGee. (2018). The Influence of Gender, and Race/Ethnicity on Advancement in Information Technology (IT). *Information and Organization*, 28(1), 1–36. doi:10.1016/j.infoandorg.2017.12.001. 3

[61] Heather Metcalf. (2010). Stuck in the Pipeline: A Critical Review of STEM Workforce Literature. *InterActions: UCLA Journal of Education and Information Studies*, 6(2), 1–20. doi:10.5070/D462000681. 2

[62] Bertrand Meyer. (2014). *Agile! The Good, the Hype and the Ugly*. Switzerland: Springer International Publishing. 15, 16

[63] Corinne A. Moss-Racusin, John F. Dovidio, Victoria L. Brescoll, Mark J. Graham, and Jo Handelsman. (2012). Science Faculty's Subtle Gender Biases Favor Male Students.

Proceedings of the National Academy of Sciences, 109(41), 16474–16479. doi:10.1073/pnas.1211286109. 14

[64] Corinne A. Moss-Racusin, Christina Sanzari, Nava Caluori, and Helena Rabasco. (2018). Gender Bias Produces Gender Gaps in STEM Engagement. *Sex Roles*, 79(11), 651–670. doi:10.1007/s11199-018-0902-z. 12, 13

[65] Corinne A. Moss-Racusin and Helena Rabasco. (2018). Reducing Gender Identity Bias through Imagined Intergroup Contact. *Journal of Applied Social Psychology*, 48(8), 457–474. doi:10.1111/jasp.12525. 14

[66] Dante P. Myers and Debra A. Major. (2017). Work–Family Balance Self-Efficacy's Relationship with STEM Commitment: Unexpected Gender Moderation. *The Career Development Quarterly*, 65(3), 264–277. doi:10.1002/cdq.12097. 8

[67] NCWIT. (2014). NCWIT Scorecard: A Report on the Status of Women in Information Technology. Retrieved from https://www.ncwit.org/resources/ncwit-scorecard-report-status-women-information-technology. 2

[68] Safiya Noble and Sarah Roberts. (2019). *Technological Elites, the Meritocracy, and Postracial Myths in Silicon Valley*. Durham: Duke University Press. doi:10.2307/j.ctv125jhp5.10. 7

[69] Colleen E. O'Connell and Karen Korabik. (2000). Sexual Harassment: The Relationship of Personal Vulnerability, Work Context, Perpetrator Status, and Type of Harassment to Outcomes. *Journal of Vocational Behavior*, 56(3), 299–329. doi:10.1006/jvbe.1999.1717. 6

[70] Scott E. Page. (2007). *The Difference: How the Power of Diversity Creates Better Groups, Firms, Schools, and Societies*. Princeton, NJ: Princeton University Press. doi:10.1515/9781400830282. 4

[71] Scott E. Page. (2017). *The Diversity Bonus: How Great Teams Pay Off in the Knowledge Economy*. Princeton, NJ: Princeton University Press. doi:10.2307/j.ctvc77c0h. 4

[72] Daphne E. Pedersen and Krista Lynn Minnotte. (2017). Workplace Climate and STEM Faculty Women's Job Burnout. *Journal of Feminist Family Therapy*, 29(1–2), 45–65. doi:10.1080/08952833.2016.1230987. 6

[73] Kim Peters, Michelle K. Ryan, and S. Alexander Haslam. (2013). Women's Occupational Motivation: The Impact of Being a Woman in a Man's World. In Susan Vinnicombe, Ronald J. Burke, Stacy Blake-Beard and Lynda L. Moore (Eds.), *Handbook of Research on Promoting Women's Careers* (pp. 162–177). doi:10.4337/9780857938961.00015. 6

[74] Ernesto Reuben, Paola Sapienza, and Luigi Zingales. (2014). How Stereotypes Impair Women's Careers in Science. *Proceedings of the National Academy of Sciences*, 111(12), 4403–4408. doi:10.1073/pnas.1314788111. 14

[75]   Frank E. Ritter, Gordon D. Baxter, and Elizabeth F. Churchill. (2014). *Foundations for Designing User-Centered Systems*. London: Springer. doi:10.1007/978-1-4471-5134-0. 4

[76]   Kristie M. Rogers and Blake E. Ashforth. (2017). Respect in Organizations: Feeling Valued as "We" and "Me". *Journal of Management*, 43(5), 1578–1608. doi:10.1177/0149206314557159. 11

[77]   Deborah Rosenfelt and Judith Stacey. (1987). Review Essay: Second Thoughts on the Second Wave. *Feminist Review*, 27(1), 77–95. doi:10.1057/fr.1987.37. 2

[78]   Allison Scott, Freada Kapor Klein, and Uriridiakoghene Onovakpuri. (2017). Tech Leavers Study - a First-of-Its-Kind Analysis of Why People Voluntarily Leave Jobs in Tech. *Kapor Center for Social Impact*. Retrieved from http://www.kaporcenter.org/wp-content/uploads/2017/08/TechLeavers2017.pdf. 1, 3, 6

[79]   Juliane Siegeris, Jörn Freiheit, and Dagmar Krefting. (2016). The Women's Bachelor Programme "Computer Science and Business Administration" and Its Consistent Orientation to Practical Training. *Procedia-Social and Behavioral Sciences*, 228, 509–514. doi:10.1016/j.sbspro.2016.07.078. 14

[80]   Caroline Simard and Shannon K. Gilmartin. (2010). Senior Technical Women: A Profile of Success. *Anita Borg Institute for Women and Technology, California*. Retrieved from http://www.wearethecity.com/wp-content/uploads/2014/12/Senior-Technical-Women-A-Profile-of-Success-March-2010.pdf. 2, 6, 7

[81]   Pankaj Singh, Damodar Suar, and Michael P. Leiter. (2012). Antecedents, Work-Related Consequences, and Buffers of Job Burnout among Indian Software Developers. *Journal of Leadership & Organizational Studies*, 19(1), 83–104. doi:10.1177/1548051811429572. 6

[82]   Romila Singh, Yejun Zhang, Min Wan, and Nadya A. Fouad. (2018). Why Do Women Engineers Leave the Engineering Profession? The Roles of Work–Family Conflict, Occupational Commitment, and Perceived Organizational Support. *Human Resource Management*, 57(4), 901–914. doi:10.1002/hrm.21900. 3

[83]   Kieran Snyder. (2014). Why Women Leave Tech: It's the Culture, Not Because 'Math Is Hard'. Retrieved from http://fortune.com/2014/10/02/women-leave-tech-culture/. 7

[84]   Olivia Solon. (2017). The Lawyers Taking on Silicon Valley Sexism: 'It's Far Worse Than People Know'. Retrieved from https://www.theguardian.com/money/2017/sep/02/lawyers-silicon-valley-sexism-worse-google-uber-lawless-sisters. 6

[85]   Sanjay Srivastava, Steve Guglielmo, and Jennifer S. Beer. (2010). Perceiving Others' Personalities: Examining the Dimensionality, Assumed Similarity to the Self, and Sta-

bility of Perceiver Effects. *Journal of Personality and Social Psychology*, 98(3), 520–534. doi:10.1037/a0017057. 4

[86]   Una Tellhed, Martin Bäckström, and Fredrik Björklund. (2016). Will I Fit in and Do Well? The Importance of Social Belongingness and Self-Efficacy for Explaining Gender Differences in Interest in STEM and Heed Majors. *Sex Roles*, 1-11. doi:10.1007/s11199-016-0694-y. 6

[87]   Josh Terrell, Andrew Kofink, Justin Middleton, Clarissa Rainear, Emerson Murphy-Hill, Chris Parnin, and Jon Stallings. (2017). Gender Differences and Bias in Open Source: Pull Request Acceptance of Women Versus Men. *PeerJ Computer Science*, 3(e111), 1–39. doi:10.7717/peerj-cs.111. 7

[88]   Andria Thomas, Joe Dougherty, Scott Strand, Abhinav Nayar, and Mariam Janani. (2016). Decoding Diversity: The Financial and Economic Returns to Diversity in Tech. *Intel Corporation; Dalberg Global Development Advisors*. Retrieved from https://newsroom. intel.com/newsroom/wp-content/uploads/sites/11/2016/07/Diversity_report_7.7.16_. 4

[89]   Bruce W. Tuckman and Mary Ann C. Jensen. (1977). Stages of Small-Group Development Revisited. *Group & Organization Studies*, 2(4), 419–427. doi:10.1177/105960117700200404. 16

[90]   Zeynep Tufekci. (2013). It's a Man's Phone. *Medium*. Retrieved from https://medium. com/technology-and-society/a26c6bee1b69. 5

[91]   Zeynep Tufekci. (2017). *Twitter and Tear Gas: The Power and Fragility of Networked Protest*. New Haven, CT: Yale University Press. 5

[92]   Ellen B. Van Oosten, Kathleen Buse, and Diana Bilimoria. (2017). The Leadership Lab for Women: Advancing and Retaining Women in STEM through Professional Development. *Frontiers in Psychology*, 8(2138), 1–5. doi:10.3389/fpsyg.2017.02138. 12

[93]   Dennis Vilorio. (2014). STEM 101: Intro to Tomorrow's Jobs. *Occupational Outlook Quarterly*, 58(1), 2–12. 1

[94]   Sara Wachter-Boettcher. (2017). *Technically Wrong: Sexist Apps, Biased Algorithms, and Other Threats of Toxic Tech*. New York, NY: WW Norton & Company. 5

[95]   Gayna Williams. (2013). *The Business of Gender: Is Your Product Gender-Neutral?* Retrieved from http://www.ifshecanican.com/uploads/4/7/8/6/4786906/isyouproduct_genderneutral-20141.pdf. 4

[96]   Joan C. Williams, Katherine W. Phillips, and Erika V. Hall. (2016). Tools for Change: Boosting the Retention of Women in the STEM Pipeline. *Journal of Research in Gender Studies*, 6(1), 11–75. doi:10.22381/JRGS6120161. 2

[97] Holly O. Witteman, Michael Hendricks, Sharon Straus, and Cara Tannenbaum. (2018). Female Grant Applicants are Equally Successful When Peer Reviewers Assess the Science, but Not When they Assess the Scientist. *The Lancet*, 393(10171), P531-540. doi:10.1016/S0140-6736(18)32611-4. 14

A DYNAMIC, VALUING TEAM THAT'S UP TO SOMETHING BIG

[1] Khaled Albusays, Pernille Bjorn, Laura Dabbish, Denae Ford, Emerson Murphy-Hill, Alexander Serebrenik, and Margaret-Anne Storey. (2021). The Diversity Crisis in Software Development. *IEEE Software*, 38(2), 19–25. doi:10.1109/MS.2020.3045817. 32

[2] Sharla Alegria. (2019). Escalator or Step Stool? Gendered Labor and Token Processes in Tech Work. *Gender & Society*, 33(5), 722–745. doi:10.1177/0891243219835737. 37

[3] Jutta Allmendinger and J. Richard Hackman. (1995). The More, the Better? A Four-Nation Study of the Inclusion of Women in Symphony Orchestras. *Social Forces*, 74(2), 423–460. doi:10.2307/2580487. 39

[4] Katerina Bezrukova, Karen A. Jehn, Elaine L. Zanutto, and Sherry M. B. Thatcher. (2009). Do Workgroup Faultlines Help or Hurt? A Moderated Model of Faultlines, Team Identification, and Group Performance. *Organization Science*, 20(1), 35–50. doi:10.1287/orsc.1080.0379. 32

[5] Iris Bohnet. (2016). *What Works: Gender Equality by Design*. Cambridge, MA: Harvard University Press. doi:10.4159/9780674545991. 33

[6] Tolga Bolukbasi, Kai-Wei Chang, James Y Zou, Venkatesh Saligrama, and Adam T Kalai. (2016). Man Is to Computer Programmer as Woman Is to Homemaker? Debiasing Word Embeddings. *Advances in Neural Information Processing Systems*, 4349–4357. 32

[7] Sapna Cheryan and Hazel Rose Markus. (2020). Masculine Defaults: Identifying and Mitigating Hidden Cultural Biases. *Psychological Review*, 127(6), 1022–1052. doi:10.1037/rev0000209. 32

[8] François Chiocchio and Hélène Essiembre. (2009). Cohesion and Performance: A Meta-Analytic Review of Disparities Between Project Teams, Production Teams, and Service Teams. *Small Group Research*, 40(4), 382–420. doi:10.1177/1046496409335103. 31

[9] Elizabeth Dwoskin and Nitasha Tiku. (2021 April 6). A Recruiter Joined Facebook to Help It Meet Its Diversity Targets. He Says Its Hiring Practices Hurt People of Color. *The Washington Post*. 40

[10]   Bruno Dyck and Frederick A. Starke. (1999). The Formation of Breakaway Organizations: Observations and a Process Model. *Administrative Science Quarterly*, 44(4), 792–822. doi:10.2307/2667056. 32

[11]   Monique Frize, Lenka Lhotska, Loredana G. Marcu, Magdalena Stoeva, Gilda Barabino, Fatimah Ibrahim, Sierin Lim, Eleni Kaldoudi, Ana Maria Marques da Silva, and Peck Ha Tan. (2021). The Impact of Covid-19 Pandemic on Gender-Related Work from Home in Stem Fields—Report of the Wimpbme Task Group. *Gender, Work & Organization*. doi:10.1111/gwao.12690. 37

[12]   Gary Garrison, Robin L. Wakefield, Xiaobo Xu, and Sang Hyun Kim. (2010). Globally Distributed Teams: The Effect of Diversity on Trust, Cohesion and Individual Performance. *ACM SIGMIS Database: The DATABASE for Advances in Information Systems*, 41(3), 27–48. doi:10.1145/1851175.1851178. 31

[13]   Lucy L. Gilson and Christina E. Shalley. (2004). A Little Creativity Goes a Long Way: An Examination of Teams' Engagement in Creative Processes. *Journal of Management*, 30(4), 453–470. doi:10.1016/j.jm.2003.07.001. 29

[14]   A. H. Gupta. (2020, April 14). It's Not Just You: In Online Meetings, Many Women Can't Get a Word In. In *The New York Times*. 37

[15]   James K. Harter, Frank L. Schmidt, and Corey L. M. Keyes. (2003). Well-Being in the Workplace and Its Relationship to Business Outcomes: A Review of The Gallup Studies. *Flourishing: Positive Psychology and the Life Well-lived*, 2, 205–224. doi:10.1037/10594-009. 27

[16]   Inga J. Jasmin Hoever, Jing Zhou, and Daan van Knippenberg. (2017). Different Strokes for Different Teams: The Contingent Effects of Positive and Negative Feedback on the Creativity of Informationally Homogeneous and Diverse Teams. *Academy of Management Journal*, 61(6), 2159–2181. doi:10.5465/amj.2016.0642. 29

[17]   Charlotte Holgersson and Laurence Romani. (2020). Tokenism Revisited: When Organizational Culture Challenges Masculine Norms, the Experience of Token Is Transformed. *European Management Review*, 17(3), 649–661. doi:10.1111/emre.12385. 39

[18]   Karen Holtzblatt and Nicola Marsden. (2018). Retaining Women in Technology—Uncovering and Measuring Key Dimensions of Daily Work Experiences. *IEEE International Conference on Engineering, Technology and Innovation (ICE/ITMC)*, 148–155. doi:10.1109/ICE.2018.8436351. 39

[19]   Astrid C. Homan, Daan Van Knippenberg, Gerben A. Van Kleef, and Carsten K. W. De Dreu. (2007). Bridging Faultlines by Valuing Diversity: Diversity Beliefs, Information

Elaboration, and Performance in Diverse Work Groups. *Journal of Applied Psychology*, 92(5), 1189–1199. doi:10.1037/0021-9010.92.5.1189. 32

[20] Lise Jans, Tom Postmes, and Karen I. Van der Zee. (2012). Sharing Differences: The Inductive Route to Social Identity Formation. *Journal of Experimental Social Psychology*, 48(5), 1145–1149. doi:10.1016/j.jesp.2012.04.013. 34

[21] Rosabeth Moss Kanter. (1977). *Men and Women of the Corporation* (Vol. 5049): Basic Books. 37, 39

[22] Etlyn J. Kenny and Rory Donnelly. (2019). Navigating the Gender Structure in Information Technology: How Does this Affect the Experiences and Behaviours of Women? *Human Relations*. doi:10.1177/0018726719828449. 32

[23] Alison M. Konrad, Vicki Kramer, and Sumru Erkut. (2008). The Impact of Three or More Women on Corporate Boards. *Organizational Dynamics*, 37(2), 145–164. doi:10.1016/j.orgdyn.2008.02.005. 39

[24] Dora C. Lau and J. Keith Murnighan. (1998). Demographic Diversity and Faultlines: The Compositional Dynamics of Organizational Groups. *Academy of Management Review*, 23(2), 325–340. doi:10.2307/259377. 32

[25] Martin Lea, Russell Spears, and Susan E. Watt. (2007). Visibility and Anonymity Effects on Attraction and Group Cohesiveness. *European Journal of Social Psychology*, 37(4), 761–773. doi:10.1002/ejsp.398. 34

[26] Jessica Lipnack and Jeffrey Stamps. (2008). *Virtual Teams: People Working Across Boundaries with Technology*. John Wiley & Sons. 35

[27] Bertolt Meyer and Carsten C. Schermuly. (2012). When Beliefs Are Not Enough: Examining the Interaction of Diversity Faultlines, Task Motivation, and Diversity Beliefs on Team Performance. *European Journal of Work and Organizational Psychology*, 21(3), 456–487. doi:10.1080/1359432X.2011.560383. 32

[28] Eric Molleman. (2005). Diversity in Demographic Characteristics, Abilities and Personality Traits: Do Faultlines Affect Team Functioning? *Group Decision and Negotiation*, 14(3), 173–193. doi:10.1007/s10726-005-6490-7. 31

[29] Ashley A. Niler, Raquel Asencio, And Leslie A. Dechurch. (2020). Solidarity in STEM: How Gender Composition Affects Women's Experience in Work Teams. *Sex Roles*, 82(3–4), 142–154. doi:10.1007/s11199-019-01046-8. 39

[30] Safiya Noble and Sarah Roberts. (2019). *Technological Elites, the Meritocracy, and Postracial Myths in Silicon Valley*. California Digital Library University of California: https://escholarship.org/uc/item/7z3629nh. doi:10.2307/j.ctv125jhp5.10. 32

[31]   Brian A. Nosek, Mahzarin R. Banaji, and Anthony G. Greenwald. (2002). Math = Male, Me = Female, Therefore Math ≠ Me. *Journal of Personality and Social Psychology*, 83(1), 44–59. doi:10.1037/0022-3514.83.1.44. 32

[32]   Judith S. Olson and Gary M. Olson. (2014). Bridging Distance: Empirical Studies of Distributed Teams. *Human-Computer Interaction and Management Information Systems: Applications. Advances in Management Information Systems* (pp. 117–134): Routledge. 37

[33]   Corinne Post. (2015). When Is Female Leadership an Advantage? Coordination Requirements, Team Cohesion, and Team Interaction Norms. *Journal of Organizational Behavior*, 36(8), 1153–1175. doi:10.1002/job.2031. 30, 31

[34]   Gary N. Powell. (2018). *Women and Men in Management*. Sage Publications. 37

[35]   Nuria Rovira-Asenjo, Agnieszka Pietraszkiewicz, Sabine Sczesny, Tània Gumí, Roger Guimerà, and Marta Sales-Pardo. (2017). Leader Evaluation and Team Cohesiveness in the Process of Team Development: A Matter of Gender? *PLoS One*. doi:10.1371/journal.pone.0186045. 30

[36]   Aspen Russell and Eitan Frachtenberg. (2021). After the Pandemic: Tech, Work, and the Tech Workforce. *IEEE Computer*, 54(7). doi:10.36227/techrxiv.13278092.v3. 35

[37]   Melanie C. Steffens, Maria Angels Viladot, and Carolin Scheifele. (2019). Male Majority, Female Majority, or Gender Diversity: How Do Proportions Affect Gender Stereotyping and Women Leaders' Well-Being? *Frontiers in Psychology*, 10, 1037. doi:10.3389/fpsyg.2019.01037. 32

[38]   Jane G. Stout and Heather M. Wright. (2016). Lesbian, Gay, Bisexual, Transgender, and Queer Students' Sense of Belonging in Computing: An Intersectional Approach. *Computing in Science & Engineering*, 18(3), 24–30. doi:10.1109/MCSE.2016.45. 32

[39]   Mariateresa Torchia, Andrea Calabrò, and Morten Huse. (2011). Women Directors on Corporate Boards: From Tokenism to Critical Mass. *Journal of Business Ethics*, 102(2), 299–317. doi:10.1007/s10551-011-0815-z. 37

[40]   Sarah F. van der Land, Alexander P. Schouten, Frans Feldberg, Marleen Huysman, and Bart van den Hooff. (2015). Does Avatar Appearance Matter? How Team Visual Similarity and Member–Avatar Similarity Influence Virtual Team Performance. *Human Communication Research*, 41(1), 128–153. doi:10.1111/hcre.12044. 35

[41]   Tanja van der Lippe and Zoltán Lippényi. (2020). Co-Workers Working from Home and Individual and Team Performance. *New Technology, Work and Employment*, 35(1), 60–79. doi:10.1111/ntwe.12153. 36

[42]   Daan Van Knippenberg and Julija N. Mell. (2016). Past, Present, and Potential Future of Team Diversity Research: From Compositional Diversity to Emergent Diversity. *Organizational Behavior and Human Decision Processes*, 136, 135–145. doi:10.1016/j.obhdp.2016.05.007. 33

[43]   Daan Van Knippenberg and Michaela C. Schippers. (2007). Work Group Diversity. *Annual Review of Psychology*, 58, 515–541. doi:10.1146/annurev.psych.58.110405.085546. 35

[44]   Daan Van Knippenberg, Wendy P. Van Ginkel, and Astrid C. Homan. (2013). Diversity Mindsets and the Performance of Diverse Teams. *Organizational Behavior and Human Decision Processes*, 121(2), 183–193. doi:10.1016/j.obhdp.2013.03.003. 40

[45]   Ruth van Veelen and Elze G. Ufkes. (2019). Teaming Up Or Down? A Multisource Study on the Role of Team Identification and Learning in the Team Diversity–Performance Link. *Group & Organization Management*, 44(1), 38–71. doi:10.1177/1059601117750532. 33

[46]   Ruth Wageman, Heidi Gardner, and Mark Mortensen. (2012). The Changing Ecology of Teams: New Directions for Teams Research. *Journal of Organizational Behavior*, 33(3), 301–315. doi:10.1002/job.1775. 29

[47]   Joseph B. Walther. (2018). The Emergence, Convergence, and Resurgence of Intergroup Communication Theory in Computer-Mediated Communication. *Atlantic Journal of Communication*, 26(2), 86–97. doi:10.1080/15456870.2018.1432616. 34

[48]   Marla Baskerville Watkins, Aneika Simmons, and Elizabeth Umphress. (2019). It's Not Black and White: Toward a Contingency Perspective on the Consequences of Being a Token. *Academy of Management Perspectives*, 33(3), 334–365. doi:10.5465/amp.2015.0154. 39

[49]   Mark G. Wilson, David M. Dejoy, Robert J. Vandenberg, Hettie A. Richardson, and Allison L. McGrath. (2004). Work Characteristics and Employee Health and Well-Being: Test of a Model of Healthy Work Organization. *Journal of Occupational and Organizational Psychology*, 77(4), 565–588. doi:10.1348/0963179042596522. 28

STIMULATING WORK

[1]   Sarah Blanchard Kyte and Catherine Riegle-Crumb. (2017). Perceptions of the Social Relevance of Science: Exploring the Implications for Gendered Patterns in Expectations of Majoring in STEM Fields. *Social Sciences*, 6(19), 1–17. doi:10.3390/socsci6010019. 44

[2]   Iris Bohnet. (2016). *What Works: Gender Equality by Design*. Cambridge, MA: Harvard University Press. doi:10.4159/9780674545991. 53

[3]   Marcus Buckingham and Ashley Goodall. (2019). The Power of Hidden Teams. *Harvard Business Review's The Big Idea*, May 14th, 2019. 46

[4]   Coleen Carrigan. (2018). 'Different Isn't Free': Gender@ Work in a Digital World. *Ethnography*, 19(3), 336-359. doi:10.1177/1466138117728737. 50, 51, 52

[5]   Emily Chang. (2019). Brotopia: Breaking up the Boys' Club of Silicon Valley: Portfolio. 51

[6]   Rob Cross, Amy Edmondson, and Wendy Murphy. (2020). A Noble Purpose Alone Won't Transform Your Company. *MIT Sloan Management Review*, 61(2), 37-43. 49

[7]   Amy Cuddy, Susan Fiske, and Peter Glick. (2008). Warmth and competence as universal dimensions of social perception: the stereotype content model and the bias map. *Advances in Experimental Social Psychology*, 40, 61–149. doi:10.1016/S0065-2601(07)00002-0. 51

[8]   Irene E. De Pater, Annelies E. M. Van Vianen, Ronald H. Humphrey, Randall G. Sleeth, Nathan S. Hartman, and Agneta H. Fischer. (2009). Individual task choice and the division of challenging tasks between men and women. *Group & Organization Management*, 34(5), 563–589. doi:10.1177/1059601108331240. 49, 52

[9]   Alice H. Eagly, Christa Nater, David I. Miller, Michèle Kaufmann, and Sabine Sczesny. (2019). Gender stereotypes have changed: a cross-temporal meta-analysis of us public opinion polls from 1946 to 2018. *American Psychologist*, 75(3), 301–315. doi:10.1037/amp0000494. 49

[10]   Naomi Ellemers. (2018). Gender stereotypes. *Annual Review of Psychology*, 69(Jan), 69:275–298. doi:10.1146/annurev-psych-122216-011719. 50

[11]   Nancy N. Heilbronner. (2013). The STEM pathway for women: what has changed? *Gifted Child Quarterly*, 57(1), 39–55. doi:10.1177/0016986212460085. 44

[12]   Madeline E Heilman, Suzette Caleo, and May Ling Halim. (2010). Just the thought of it!: effects of anticipating computer-mediated communication on gender stereotyping. *Journal of Experimental Social Psychology*, 46(4), 672–675. doi:10.1016/j.jesp.2010.02.005. 52

[13]   Madeline E. Heilman. (2012). Gender stereotypes and workplace bias. *Research in Organizational Behavior*, 32, 113–135. doi:10.1016/j.riob.2012.11.003. 51

[14]   Madeline E. Heilman, Francesca Manzi, and Susanne Braun. (2015). *Presumed Incompetent: Perceived Lack of Fit and Gender Bias in Recruitment and Selection* (Vol. 90). Cheltenham, UK: Edward Elgar. doi:10.4337/9781782547709.00014. 50

[15]   Susan C. Herring and Sharon Stoerger. (2013). Gender and (a) Nonymity in Computer-Mediated Communication. In J. Holmes, M. Meyerhoff, and S. Ehrlich (Eds.), *The*

Handbook of Language, Gender, and Sexuality (pp. 567–586). Hoboken, NJ: Wiley-Blackwell Publishing. doi:10.1002/9781118584248.ch29. 52

[16] Jo Hutchinson. (2014). 'Girls into STEM and Komm Mach Mint': English and German Approaches to Support Girls' Stem Career-Related Learning. *Journal of the National Institute for Career Education and Counselling*, 32(1), 27–34. 50

[17] Seulki Jang, Tammy D. Allen, and Joseph Regina. (2020). Office Housework, Burnout, and Promotion: Does Gender Matter? *Journal of Business and Psychology*, 36, 793–805. doi:10.1007/s10869-020-09706-3. 54

[18] W. Brad Johnson and David Smith. (2016). *Athena Rising: How and Why Men Should Mentor Women*: Routledge. doi:10.4324/9781315213163. 54

[19] Eden B. King, Whitney Botsford, Michelle R. Hebl, Stephanie Kazama, Jeremy F. Dawson, and Andrew Perkins. (2012). Benevolent Sexism at Work: Gender Differences in the Distribution of Challenging Developmental Experiences. *Journal of Management*, 38(6), 1835–1866. doi:10.1177/0149206310365902. 58

[20] Nicola Marsden and Monika Pröbster. (2019). Personas and Identity - Looking at Multiple Identities to Inform the Construction of Personas. *Proceedings of the SIGCHI Conference on Human Factors in Computing Systems (CHI '19)*, Paper 335, 1–14. doi:10.1145/3290605.3300565. 52

[21] Muriel Niederle and Alexandra H. Yestrumskas. (2008). Gender Differences in Seeking Challenges: The Role of Institutions. *National Bureau of Economic Research*, 13922. doi:10.3386/w13922. 55

[22] Elena Prieto-Rodriguez, Kristina Sincock, and Karen Blackmore. (2020). STEM Matter: Results from a Systematic Review of Secondary School Interventions for Girls. *International Journal of Science Education*, 42(7), 1144–1161. doi:10.1080/09500693.2020.1749909. 50

[23] Allison Scott, Freada Kapor Klein, and Uriridiakoghene Onovakpuri. (2017). *Tech Leavers Study - a First-of-Its-Kind Analysis of Why People Voluntarily Leave Jobs in Tech: Kapor Center for Social Impact*. 52

[24] Carroll Seron, Susan Silbey, Erin Cech, and Brian Rubineau. (2018). "I am Not a Feminist, But...": Hegemony of a Meritocratic Ideology and the Limits of Critique Among Women in Engineering. *Work and Occupations*, 45(2), 131–167. doi:10.1177/0730888418759774. 51

[25] Simon Sinek. (2009). *Start with Why: How Great Leaders Inspire Everyone to Take Action*: Penguin. 49

[26]   Louise Soe and Elaine K. Yakura. (2008). What's Wrong with the Pipeline? As-
         sumptions About Gender and Culture in It Work. *Women's Studies*, 37(3), 176–201.
         doi:10.1080/00497870801917028. 49, 52

[27]   Divy Thakkar, Nithya Sambasivan, Purva Kulkarni, Pratap Kalenahalli Sudarshan, and
         Kentaro Toyama. (2018). The Unexpected Entry and Exodus of Women in Computing
         and H C I in India. *Proceedings of the 2018 CHI Conference on Human Factors in Comput-
         ing Systems*, 352, 1–12. doi:10.1145/3173574.3173926. 50

[28]   Sarah Thébaud and Maria Charles. (2018). Segregation, Stereotypes, and STEM. *Social
         Sciences*, 7(7), 111. doi:10.3390/socsci7070111. 50

[29]   Roli Varma. (2019). Women in Computing Education. In C. Frieze and J.L. Quesenberry
         (Eds.), *Cracking the Digital Ceiling: Women in Computing Around the World* (p. 299): Cam-
         bridge University Press. doi:10.1017/9781108609081.018. 50

[30]   Joseph B. Walther. (2012). Interaction Through Technological Lenses: Computer-Me-
         diated Communication and Language. *Journal of Language and Social Psychology*, 31(4),
         397–414. doi:10.1177/0261927x12446610. 52

[31]   Joan C. Williams. (2014). Hacking tech's diversity problem. *Harvard Business Review*,
         92(10), 94–100. 54

[32]   Joan C. Williams and Rachel Dempsey. (2018). *What Works for Women at Work: Four Pat-
         terns Working Women Need to Know*: NYU Press. 52

THE PUSH AND SUPPORT

[1]   James Andreoni and Lise Vesterlund. (2001). Which Is the Fair Sex? Gender
         Differences in Altruism. *The Quarterly Journal of Economics*, 116(1), 293–312.
         doi:10.1162/003355301556419. 59

[2]   Yuen Lam Bavik, Jason D. Shaw, and Xiao-Hua Wang. (2020). Social Support: Multidis-
         ciplinary Review, Synthesis, and Future Agenda. *Academy of Management Annals*, 14(2),
         726–758. doi:10.5465/annals.2016.0148. 72

[3]   Alison Booth, Lina Cardona-Sosa, and Patrick Nolen. (2014). Gender Differences in
         Risk Aversion: Do Single-Sex Environments Affect Their Development? *Journal of Eco-
         nomic Behavior & Organization*, 99, 126–154. doi:10.1016/j.jebo.2013.12.017. 59

[4]   Prisca Brosi, Matthias Spörrle, Isabell M. Welpe, and Madeline E. Heilman. (2016).
         Expressing Pride: Effects on Perceived Agency, Communality, and Stereotype-Based
         Gender Disparities. *Journal of Applied Psychology*, 101(9), 1319–1328. doi:10.1037/
         apl0000122. 64

[5]   Hancheng Cao, Chia-Jung Lee, Shamsi Iqbal, Mary Czerwinski, Priscilla N. Y. Wong, Sean Rintel, Brent Hecht, Jaime Teevan, and Longqi Yang. (2021). Large Scale Analysis of Multitasking Behavior During Remote Meetings. *Proceedings of the 2021 CHI Conference on Human Factors in Computing Systems*, 448, 1–13. doi:10.1145/3411764.3445243. 73

[6]   Dan S. Chiaburu and David A. Harrison. (2008). Do Peers Make The Place? Conceptual Synthesis and Meta-Analysis of Coworker Effects On Perceptions, Attitudes, OCBs, and Performance. *Journal of Applied Psychology*, 93(5), 1082–1103. doi:10.1037/0021-9010.93.5.1082. 66

[7]   Susan Davis-Ali. (2017). *Advancing Women Technologists into Positions of Leadership*. Anita Borg Institute. 66

[8]   Anna-Lena Dicke, Nayssan Safavian, and Jacquelynne S. Eccles. (2019). Traditional Gender Role Beliefs and Career Attainment in STEM: A Gendered Story? *Frontiers in Psychology*, 10(1053), 1–14. doi:10.3389/fpsyg.2019.01053. 61

[9]   Catherine C. Eckel, Lata Gangadharan, Philip Johnson Grossman, and Nina Xue. (2021). The Gender Leadership Gap: Insights from Experiments. In Ananish Chaudhuri (Ed.), *A Research Agenda for Experimental Economics* (pp. 137–162). Northampton, MA: Edward Elgar Publishing. doi:10.4337/9781789909852.00014. 59

[10]   Merideth Ferguson, Dawn Carlson, K. Michele Kacmar, and Jonathon R. B. Halbesleben. (2016). The Supportive Spouse at Work: Does Being Work-Linked Help? *Journal of Occupational Health Psychology*, 21(1), 37–50. doi:10.1037/a0039538. 66

[11]   Susan Fiske, Amy Cuddy, and Peter Glick. (2007). Universal Dimensions of Social Cognition: Warmth and Competence. *Trends in Cognitive Sciences*, 11(2), 77–83. doi:10.1016/j.tics.2006.11.005. 64

[12]   Susan T. Fiske and Laura E. Stevens. (1993). *What's So Special about Sex? Gender Stereotyping and Discrimination*. Sage Publications, Inc. 64, 65

[13]   Susan T. Fiske, Juan Xu, Amy C. Cuddy, and Peter Glick. (1999). (Dis) Respecting Versus (Dis) Liking: Status and Interdependence Predict Ambivalent Stereotypes of Competence and Warmth. *Journal of Social Issues*, 55(3), 473–489. doi:10.1111/0022-4537.00128. 64

[14]   Andreas Friedl, Andreas Pondorfer, and Ulrich Schmidt. (2020). Gender Differences in Social Risk Taking. *Journal of Economic Psychology*, 77(102182), 1–17. doi:10.1016/j.joep.2019.06.005. 58

[15]   Madeline E. Heilman and Julie J. Chen. (2005). Same Behavior, Different Consequences: Reactions to Men's and Women's Altruistic Citizenship Behavior. *Journal of Applied Psychology*, 90(3), 431–441. doi:10.1037/0021-9010.90.3.431. 64

[16]   Madeline E. Heilman and Tyler G. Okimoto. (2007). Why Are Women Penalized for Success at Male Tasks?: The Implied Communality Deficit. *Journal of Applied Psychology*, 92(1), 81–92. doi:10.1037/0021-9010.92.1.81. 64

[17]   Ivona Hideg and Winny Shen. (2019). Why Still so Few? A Theoretical Model of The Role of Benevolent Sexism and Career Support in the Continued Underrepresentation of Women in Leadership Positions. *Journal of Leadership & Organizational Studies*, 26(3), 287–303. doi:10.1177/1548051819849006. 66

[18]   Katherine A. Karl, Joy V. Peluchette, and Navid Aghakhani. (2021). Virtual Work Meetings During the Covid-19 Pandemic: The Good, Bad, and Ugly. *Small Group Research*. doi:10.1177/10464964211015286. 72

[19]   Eden B. King, Whitney Botsford, Michelle R. Hebl, Stephanie Kazama, Jeremy F. Dawson, and Andrew Perkins. (2012). Benevolent Sexism at Work: Gender Differences in the Distribution of Challenging Developmental Experiences. *Journal of Management*, 38(6), 1835–1866. doi:10.1177/0149206310365902. 59

[20]   Gamze Koseoglu, Terry C. Blum, and Christina E. Shalley. (2018). Gender Similarity, Coworker Support, and Job Attitudes. *Journal of Management & Organization*, 26(5), 880–898. doi:10.1017/jmo.2018.40. 66

[21]   Anastasia Kuzminykh and Sean Rintel. (2020). Low Engagement as a Deliberate Practice of Remote Participants in Video Meetings. *Extended Abstracts of the 2020 CHI Conference on Human Factors in Computing Systems*, 1–9. doi:10.1145/3334480.3383080. 72

[22]   Emily R. Mondschein, Karen E. Adolph, and Catherine S. Tamis-LeMonda. (2000). Gender Bias in Mothers' Expectations about Infant Crawling. *Journal of Experimental Child Psychology*, 77(4), 304–316. doi:10.1006/jecp.2000.2597. 59

[23]   Jessica Nordell. (2021). *The End of Bias: A Beginning*. Metropolitan Books. 65

[24]   Belle Rose Ragins, John L. Cotton, and Janice S. Miller. (2000). Marginal Mentoring: The Effects of Type of Mentor, Quality of Relationship, and Program Design on Work and Career Attitudes. *Academy of Management Journal*, 43(6), 11771–194. doi:10.5465/1556344. 73

[25]   Vincent Rousseau and Caroline Aubé. (2010). Social Support at Work and Affective Commitment to the Organization: The Moderating Effect of Job Resource Ad-

equacy and Ambient Conditions. *The Journal of Social Psychology*, 150(4), 321–340. doi:10.1080/00224540903365380. 58, 73

[26] Laurie A. Rudman and Peter Glick. (2001). Prescriptive Gender Stereotypes and Backlash Toward Agentic Women. *Journal of Social Issues*, 57(4), 743–762. doi:10.1111/0022-4537.00239. 64

[27] Laurie A. Rudman. (1998). Self-Promotion as a Risk Factor for Women: The Costs and Benefits of Counterstereotypical Impression Management. *Journal of Personality and Social Psychology*, 74(3), 629–645. doi:10.1037/0022-3514.74.3.629. 64

[28] Laurie A. Rudman and Julie E. Phelan. (2008). Backlash Effects for Disconfirming Gender Stereotypes in Organizations. *Research in Organizational Behavior*, 28, 61–79. doi:10.1016/j.riob.2008.04.003. 64

[29] Barbara Schwarze and Judith Elisabeth Bräuer. (2019). Erfolgsfaktoren in der Studien- und Berufsorientierung MINT. Das Beispiel "Niedersachsen-Technikum." In Yvonne Haffner and Lena Loge (Eds.), *Frauen in Technik und Naturwissenschaft: Eine Frage der Passung: Aktuelle Erkenntnisse und Einblicke in Orientierungsprojekte* (pp. 183–209). Verlag Barbara Budrich. doi:10.2307/j.ctvdf0mqh.11. 59

[30] Jared Spataro. (2020). How Remote Work Impacts Collaboration: Findings from our Team. Retrieved from https://www.microsoft.com/en-us/microsoft-365/blog/2020/04/22/how-remote-work-impacts-collaboration-findings-team/. 72

[31] Rosemary Stockdale and Therese Keane. (2016). Influencing the Influencers: The Role of Mothers in IT Career Choices. *Journal of Information Technology Education: Innovations in Practice*, 15, 181–194. doi:10.28945/3624. 59

[32] Pamela S. Tolbert, Mary E. Graham, and Alice O. Andrews. (1999). Group Gender Composition and Work Group Relations: Theories, Evidence, and Issues. Retrieved from https://hdl.handle.net/1813/75618. 65

[33] Julio Mancuso Tradenta, Ananta Neelim, and Joseph Vecci. (2019). Gender Differences in Self-Promotion: Understanding the Female Modesty Constraint. Available at *SSRN* 3039233, 1-40. doi:10.2139/ssrn.3039233. 64

[34] Eileen M. Trauth, Jeria L. Quesenberry, and Haijan Huang. (2008). A Multicultural Analysis of Factors Influencing Career Choice for Women in the Information Technology Workforce. *Journal of Global Information Management*, 16(4), 1–23. doi:10.4018/jgim.2008100101.

[35]   Eileen M. Trauth, Jeria L. Quesenberry, and Haiyan Huang. (2009). Retaining Women in the U.S. IT Workforce: Theorizing the Influence of Organizational Factors. *European Journal of Information Systems*, 18(5), 476-497. doi:10.1057/ejis.2009.31. 59

[36]   Sandra V. Turner, Phyllis W. Bernt, and Norma Pecora. (2002). Why Women Choose Information Technology Careers: Educational, Social, and Familial Influences. Retrieved from https://eric.ed.gov/?id=ED465878. 59

[37]   Bin Wang, Yukun Liu, Jing Qian, and Sharon K. Parker. (2021). Achieving Effective Remote Working During the Covid-19 Pandemic: A Work Design Perspective. *Applied Psychology*, 70(1), 16–59. doi:10.1111/apps.12290. 72

[38]   Joan C. Williams. (2014). Hacking Tech's Diversity Problem. *Harvard Business Review*, 92(10), 94–100. 65

LOCAL ROLE MODELS

[1]   Tammy D. Allen, Lillian T. Eby, Mark L. Poteet, Elizabeth Lentz, and Lizzette Lima. (2004). Career Benefits Associated with Mentoring for Protégés: A Meta-Analysis. *Journal of Applied Psychology*, 89(1), 127–136. doi:10.1037/0021-9010.89.1.127. 83

[2]   Sapna Cheryan, Victoria C. Plaut, Paul G. Davies, and Claude M. Steele. (2009). Ambient Belonging: How Stereotypical Cues Impact Gender Participation In Computer Science. *Journal of Personality and Social Psychology*, 97(6), 1045–1060. doi:10.1037/a0016239. 76

[3]   Sapna Cheryan, John Oliver Siy, Marissa Vichayapai, Benjamin J. Drury, and Saenam Kim. (2011). Do Female and Male Role Models who Embody Stem Stereotypes Hinder Women's Anticipated Success in STEM? *Social Psychological and Personality Science*, 2(6), 656–664. doi:10.1177/1948550611405218. 77

[4]   Nilanjana Dasgupta. (2011). Ingroup Experts and Peers as Social Vaccines who Inoculate the Self-Concept: The Stereotype Inoculation Model. *Psychological Inquiry*, 22(4), 231–246. doi:10.1080/1047840X.2011.607313. 77

[5]   Katherine Dashper. (2017). Challenging the Gendered Rhetoric Of Success? The Limitations of Women-Only Mentoring for Tackling Gender Inequality in the Workplace. *Gender, Work & Organization*, 26(4), 1–17. doi:10.1111/gwao.12262. 77

[6]   Rene F. Kizilcec, Andrew Saltarelli, Petra Bonfert-Taylor, N. H. Hanover, Michael Goudzwaard, Ella Hamonic, and Rémi Sharrock. (2020). Welcome to the Course: Early Social Cues Influence Women's Persistence in Computer Science. *Proceedings of the 2020 CHI*

Conference on Human Factors in Computing Systems, 1–13. doi:10.1145/3313831.3376752. 76

[7]   Jennifer LaCosse, Denise Sekaquaptewa, and Jill Bennett. (2016). STEM Stereotypic Attribution Bias among Women in Aan Unwelcoming Science Setting. *Psychology of Women Quarterly*, 40(3), 378–397. doi:10.1177/0361684316630965. 76

[8]   Ma Carolina Saffie-Robertson. (2020). It's Not You, it's Me: An Exploration of Mentoring Experiences for Women in STEM. *Sex Roles*, 83, 566-579. doi:10.1007/s11199-020-01129-x. 86

[9]   Louise Soe and Elaine K. Yakura. (2008). What's Wrong with the Pipeline? Assumptions about Gender and Culture in IT Work. *Women's Studies*, 37(3), 176–201. doi:10.1080/00497870801917028. 77

NONJUDGMENTAL FLEXIBILITY FOR FAMILY COMMITMENTS

[1]   Catherine Ashcraft, Brad McLain, and Elizabeth Eger. (2016). NCWIT - Women in Tech: The Facts. Retrieved from https://www.ncwit.org/sites/default/files/resources/ncwit_women-in-it_2016-full-report_final-web06012016.pdf. 93

[2]   Nino Bariola and Caitlyn Collins. (2021). The Gendered Politics of Pandemic Relief: Labor and Family Policies in Denmark, Germany, and The United States During Covid-19. *American Behavioral Scientist*, 65(12), 1671–1697. doi:10.1177/00027642211003140. 95

[3]   Mary Blair-Loy. (2009). *Competing Devotions: Career and Family among Women Executives*. Harvard University Press. 93, 99

[4]   Erin A. Cech and Mary Blair-Loy. (2019). The Changing Career Trajectories of New Parents in STEM. *Proceedings of the National Academy of Sciences*, 116(10), 4182–4187. doi:10.1073/pnas.1810862116. 93

[5]   Heejung Chung and Tanja Van der Lippe. (2018). Flexible Working, Work–Life Balance, and Gender Equality: Introduction. *Social Indicators Research*, 1–17. doi:10.1007/s11205-018-2025-x. 94

[6]   Jeffrey R. Cohen and Louise E. Single. (2001). An Examination of the Perceived Impact of Flexible Work Arrangements on Professional Opportunities in Public Accounting. *Journal of Business Ethics*, 32(4), 317–328. doi:10.1023/A:1010767521662. 99

[7]   Caitlyn Collins. (2018). The Promise and Limits of Work-Family Supports in a Shifting Policy Landscape: A Double Bind for Working Mothers in Western Germany. In Mi-

chelle Y. Janning (Ed.), *Contemporary Parenting and Parenthood: From News Headlines to New Research* (pp. 141–167). Santa Barbara, CA: Praeger Press. 94

[8] Caitlyn Collins, Liana Christin Landivar, Leah Ruppanner, and William J. Scarborough. (2020). Covid-19 and the Gender Gap in Work Hours. *Gender, Work & Organization*, 28(S1), 101–112. doi:10.1111/gwao.12506. 92

[9] Shelley J. Correll, Stephen Benard, and In Paik. (2007). Getting a Job: Is there a Motherhood Penalty? *American Journal of Sociology*, 112(5), 1297–1339. doi:10.1086/511799. 99

[10] Helen Delaney, Senior Lecturer, and Katie R. Sullivan. (2021). The Political Is Personal: Postfeminism and the Construction of the Ideal Working Mother. *Gender, Work & Organization*, 28(4), 1697–1710. doi:10.1111/gwao.12702. 99

[11] Catherine Doren. (2019). Which Mothers Pay a Higher Price? Education Differences in Motherhood Wage Penalties by Parity and Fertility Timing. *Sociological Science*, 6, 684–709. doi:10.15195/v6.a26. 99

[12] Kimberly Earles. (2020). *The Gender Divide in the Tech Sector*. The Washington State Labor Education and Research Center. 99

[13] Sylvia Fuller and C. Elizabeth Hirsh. (2019). "Family-Friendly" Jobs and Motherhood Pay Penalties: The Impact of Flexible Work Arrangements across the Educational Spectrum. *Work and Occupations*, 46(1), 3–44. doi:10.1177/0730888418771116. 99

[14] Rebecca Glauber. (2018). Trends in the Motherhood Wage Penalty and Fatherhood Wage Premium for Low, Middle, and High Earners. *Demography*, 55(5), 1663–1680. doi:10.1007/s13524-018-0712-5. 100

[15] Amrita Hari. (2017). Who Gets to 'Work Hard, Play Hard'? Gendering the Work–Life Balance Rhetoric in Canadian Tech Companies. *Gender, Work & Organization*, 24(2), 99–114. doi:10.1111/gwao.12146. 95

[16] Lena Hipp. (2019). Do Hiring Practices Penalize Women and Benefit Men for Having Children? Experimental Evidence from Germany. *European Sociological Review*, 36(2), 250–264. doi:10.1093/esr/jcz056. 99

[17] Erin Kramer Holmes, Jocelyn Wikle, Clare R. Thomas, McKell A. Jorgensen, and Braquel R. Egginton. (2020). Social Contact, Time Alone, and Parental Subjective Well-Being: A Focus on Stay-At-Home Fathers Using the American Time Use Survey. *Psychology of Men & Masculinities*, 22(3), 488–499. doi:10.1037/men0000321. 100

[18]   Kendall Houghton. (2020). Childcare and the New Part-Time: Gender Gaps in Long-Hour Professions. Retrieved from https://appam.confex.com/appam/2020/mediafile/ExtendedAbstract/Paper36154/c42e3b_9fa4ef04b11a44db81db4f1acebfb927.pdf. 100

[19]   Aimzhan Iztayeva. (2021). Custodial Single Fathers Before and During the Covid-19 Crisis: Work, Care, and Well-Being. *Social Sciences*, 10(94), 1–23. doi:10.3390/socsci10030094. 100

[20]   Shawna J. Lee, Joyce Y. Lee, and Olivia D. Chang. (2020). The Characteristics and Lived Experiences of Modern Stay-At-Home Fathers. In H.E. Fitzgerald, K. von Klitzing, N.J. Cabrera, J. Scarano de Mendonça, and T. Skjøthaug (Eds.), *Handbook of Fathers and Child Development* (pp. 537–549). Cham: Springer. 100

[21]   Lisa M. Leslie, Colleen Flaherty Manchester, Tae-Youn Park, and Si Ahn Mehng. (2012). Flexible Work Practices: A Source of Career Premiums or Penalties? *Academy of Management Journal*, 55(6), 1407–1428. doi:10.5465/amj.2010.0651. 99

[22]   Melissa Mazmanian and Ingrid Erickson. (2014). The Product of Availability: Understanding the Economic Underpinnings of Constant Connectivity. *Proceedings of the SIGCHI Conference on Human Factors in Computing Systems*, 763–772. doi:10.1145/2556288.2557381. 93

[23]   Krista Lynn Minnotte and Michael C. Minnotte. (2021). The Ideal Worker Norm and Workplace Social Support among U.S. Workers. *Sociological Focus*, 54(2), 120–137. doi:10.1080/00380237.2021.1894622. 93

[24]   Lauren D. Murphy, Candice L. Thomas, Haley R. Cobb, and Alexius E. Hartman. (2021). A R of the Lgbtq+ Work–Family Interface: What Do We Know and Where Do We Go from Here? *Journal of Organizational Behavior*, 42(2), 139–161. doi:10.1002/job.2492. 100

[25]   Irene Padavic, Robin J. Ely, and Erin M. Reid. (2020). Explaining the Persistence of Gender Inequality: The Work–Family Narrative as a Social Defense Against the 24/7 Work Culture. *Administrative Science Quarterly*, 65(1), 61–111. doi:10.1177/0001839219832310. 95, 100

[26]   Richard J. Petts, Daniel L. Carlson, and Joanna R. Pepin. (2020). A Gendered Pandemic: Childcare, Homeschooling, and Parents' Employment During Covid-19. *Gender, Work & Organization*, 28(S2), 515–534. doi:10.1111/gwao.12614. 92

[27]   Laurie A. Rudman and Kris Mescher. (2013). Penalizing Men Who Request a Family Leave: Is Flexibility Stigma a Femininity Stigma? *Journal of Social Issues*, 69(2), 322–340. doi:10.1111/josi.12017. 99, 100

[28]   Trustradius. (2021). Women in Tech Report. Retrieved from https://www.trustradius.com/buyer-blog/women-in-tech-report. 92

[29]   Catherine J. Turco. (2010). Cultural Foundations of Tokenism: Evidence from the Leveraged Buyout Industry. *American Sociological Review*, 75(6), 894–913. doi:10.1177/0003122410388491. 93

[30]   Amy S. Wharton, Sarah Chivers, and Mary Blair-Loy. (2008). Use of Formal and Informal Work–Family Policies on the Digital Assembly Line. *Work and Occupations*, 35(3), 327–350. doi:10.1177/0730888408316393. 99

[31]   Joan C. Williams, Mary Blair-Loy, and Jennifer L. Berdahl. (2013). Cultural Schemas, Social Class, and the Flexibility Stigma. *Journal of Social Issues*, 69(2), 209–234. doi:10.1111/josi.12012. 99

[32]   Alison T. Wynn and Aliya Hamid Rao. (2020). Failures of Flexibility: How Perceived Control Motivates the Individualization of Work–Life Conflict. *ILR Review*, 73(1), 61–90. doi:10.1177/0019793919848426. 98

[33]   Jill Yavorsky, George Christopher Banks, and Alyssa McGonagle. (2019). What Men Can Do to Reduce Gender Inequality in Science, Medicine, and Global Health: Small Wins and Organizational Change. *OSF Preprints*, 1–16. doi:10.31219/osf.io/pjtuy. 94, 100

PERSONAL POWER

[1]   Nancy K Baym. (2015). *Personal Connections in the Digital Age* (2nd ed.). Cambridge, UK: Polity. 113

[2]   Wiebke Bleidorn, Ruben C. Arslan, Jaap J. A. Denissen, Peter J. Rentfrow, Jochen E. Gebauer, Jeff Potter, and Samuel D. Gosling. (2016). Age and Gender Differences In Self-Esteem—A Cross-Cultural Window. *Journal of Personality and Social Psychology*, 111(3), 396–410. doi:10.1037/pspp0000078. 106

[3]   Sapna Cheryan and Hazel Rose Markus. (2020). Masculine Defaults: Identifying and Mitigating Hidden Cultural Biases. *Psychological Review*, 127(6), 1022–1052. doi:10.1037/rev0000209. 107

[4]   Pauline Rose Clance and Suzanne Ament Imes. (1978). The imposter phenomenon in high achieving women: dynamics and therapeutic intervention. *Psychotherapy: Theory, Research & Practice*, 15(3), 241–247. doi:10.1037/h0086006. 105

[5]   Laurie L. Cohen and Janet K. Swim. (1995). The differential impact of gender ratios on women and men: tokenism, self-confidence, and expectations. *Personality and Social Psychology Bulletin*, 21(9), 876–884. doi:10.1177/0146167295219001. 107

[6]   Julia T. Crawford. (2021). Imposter Syndrome for Women in Male Dominated Careers. *Hastings Women's Law Journal*, 32(2), 1–50. 105

[7]   Dario Cvencek, Ružica Brečić, Dora Gaćeša, and Andrew N. Meltzoff. (2021). Development of Math Attitudes and Math Self-Concepts: Gender Differences, Implicit–Explicit Dissociations, and Relations to Math Achievement. *Child Development*, 92(5), e940–e956. doi:10.1111/cdev.13523. 106

[8]   Nicole M. Else-Quest, Janet Shibley Hyde, and Marcia C. Linn. (2010). Cross-National Patterns of Gender Differences in Mathematics: A Meta-Analysis. *Psychological Bulletin*, 136(1), 103–127. doi:10.1037/a0018053. 106

[9]   Dulini Fernando, Laurie Cohen, and Joanne Duberley. (2018). Navigating Sexualised Visibility: A Study of British Women Engineers. *Journal of Vocational Behavior*, 113, 6–19. doi:10.1016/j.jvb.2018.06.001. 107

[10]   Laura Guillén, Margarita Mayo, and Natalia Karelaia. (2018). Appearing Self-Confident and Getting Credit for It: Why It May Be Easier for Men than Women to Gain Influence at Work. *Human Resource Management*, 57(4), 839–854. doi:10.1002/hrm.21857. 106

[11]   Ellen Lenney. (1977). Women's Self-Confidence in Achievement Settings. *Psychological Bulletin*, 84(1), 1–13. doi:10.1037/0033-2909.84.1.1. 106

[12]   Sara M. Lindberg, Janet Shibley Hyde, Jennifer L. Petersen, and Marcia C. Linn. (2010). New Trends in Gender and Mathematics Performance: A Meta-Analysis. *Psychological Bulletin*, 136(6), 1123–1135. doi:10.1037/a0021276. 106

[13]   Ashley E. Martin and Katherine W. Phillips. (2017). What "Blindness" to Gender Differences Helps Women See and Do: Implications for Confidence, Agency, and Action in Male-Dominated Environments. *Organizational Behavior and Human Decision Processes*, 142, 28–44. doi:10.1016/j.obhdp.2017.07.004. 107

[14]   Safiya Noble and Sarah Roberts. (2019). *Technological Elites, the Meritocracy, and Postracial Myths in Silicon Valley*. Durham, NC: Duke University Press. doi:10.2307/j.ctv-125jhp5.10. 107

[15]   Dinara Tokbaeva and Leona Achtenhagen. (2021). Career Resilience of Female Professionals in the Male-Dominated IT Industry in Sweden: Toward a Process Perspective. *Gender, Work & Organization*, 1–40. doi:10.1111/gwao.12671. 109

[16]   Eric Luis Uhlmann and Geoffrey L. Cohen. (2007). "I Think It, Therefore It's True": Effects of Self-Perceived Objectivity on Hiring Discrimination. *Organizational Behavior and Human Decision Processes*, 104(2), 207–223. doi:10.1016/j.obhdp.2007.07.001. 108

[17]   Joan C. Williams, Katherine W. Phillips, and Erika V. Hall. (2016). Tools for Change: Boosting the Retention of Women in the STEM Pipeline. *Journal of Research in Gender Studies*, 6(1), 11–75. doi:10.22381/JRGS6120161. 107

TEAM ONBOARDING

[1]   Talya N. Bauer. (2010). Onboarding New Employees: Maximizing Success. *SHRM Foundation's Effective Practice Guidelines Series*. 125

[2]   Talya N. Bauer. (2013). Onboarding: The Power of Connection. *Onboarding White Paper Series. SuccessFactors*. 125

[3]   Talya N. Bauer, Todd Bodner, Berrin Erdogan, Donald M. Truxillo, and Jennifer S. Tucker. (2007). Newcomer Adjustment During Organizational Socialization: A Meta-Analytic Review of Antecedents, Outcomes, and Methods. *Journal of Applied Psychology*, 92(3), 707–721. doi:10.1037/0021-9010.92.3.707. 125

[4]   Ricardo Britto, Daniela S. Cruzes, Darja Smite, and Aivars Sablis. (2018). Onboarding Software Developers and Teams in Three Globally Distributed Legacy Projects: A Multi-Case Study. *Journal of Software: Evolution and Process*, 30(4), e1921. doi:10.1002/smr.1921. 129

[5]   Rob Cross, Amy Edmondson, and Wendy Murphy. (2020). A Noble Purpose Alone Won't Transform Your Company. *MIT Sloan Management Review*, 61(2), 37–43. 125

[6]   Allison M. Ellis, Sushil S. Nifadkar, Talya N. Bauer, and Berrin Erdogan. (2017). Your New Hires Won't Succeed Unless You Onboard Them Properly. *Harvard Business Review Digital Articles*, June 20, 1–2. doi:10.5465/AMBPP.2017.282. 125

[7]   Karen Holtzblatt. (2021). The Team Onboarding Checklist. Retrieved from https://www.witops.org/the-team-onboarding-checklist/. 139

[8]   Karen Holtzblatt and Carol Farnsworth. (2018). Team Onboarding: Key Success Factors and How to Apply Them. *Presentation at Grace Hopper Celebration*. 126

[9]   John Kammeyer-Mueller, Connie Wanberg, Alex Rubenstein, and Zhaoli Song. (2013). Support, Undermining, and Newcomer Socialization: Fitting in During the First 90 Days. *Academy of Management Journal*, 56(4), 1104–1124. doi:10.5465/amj.2010.0791. 125

[10]   Paige Rodeghero, Thomas Zimmermann, Brian Houck, and Denae Ford. (2020). Please Turn Your Cameras On: Remote Onboarding of Software Developers During a Pandemic. *IEEE/ACM 43rd International Conference on Software Engineering: Software Engineering in Practice (ICSE-SEIP)*. doi:10.1109/ICSE-SEIP52600.2021.00013. 129

[11]   Gaurav G. Sharma and Klaas-Jan Stol. (2020). Exploring Onboarding Success, Organizational Fit, and Turnover Intention of Software Professionals. *Journal of Systems and Software*, 159, 1–19. doi:10.1016/j.jss.2019.110442. 125

THE CRITIQUE MEETING

[1]   Joseph A. Allen, Nale Lehmann-Willenbrock, and Steven G. Rogelberg. (2015). *The Cambridge Handbook of Meeting Science*. Cambridge University Press. doi:10.1017/CBO9781107589735. 152

[2]   Julia B. Bear, Lily Cushenbery, Manuel London, and Gary D. Sherman. (2017). Performance Feedback, Power Retention, and the Gender Gap in Leadership. *The Leadership Quarterly*, 28(6), 721–740. doi:10.1016/j.leaqua.2017.02.003. 148

[3]   Elizabeth Boling and Colin M. Gray. (2021). Instructional Design and User Experience Design: Values and Perspectives Examined through Artifact Analysis. In Brad Hokanson, Marisa Exter, Amy Grincewicz, Matthew Schmidt, and Andrew A. Tawfik (Eds.), *Intersections across Disciplines* (pp. 93–107): Springer. doi:10.1007/978-3-030-53875-0_8. 148

[4]   Amiangshu Bosu, Jeffrey C. Carver, Christian Bird, Jonathan Orbeck, and Christopher Chockley. (2017). Process Aspects and Social Dynamics of Contemporary Code Review: Insights from Open Source Development and Industrial Practice at Microsoft. *IEEE Transactions on Software Engineering*, 43(1), 56–75. doi:10.1109/tse.2016.2576451. 148

[5]   Tore Dybå, Neil Maiden, and Robert Glass. (2014). The Reflective Software Engineer: Reflective Practice. *IEEE Software*, 31(4), 32–36. doi:10.1109/MS.2014.97. 147

[6]   Michael Fagan. (2002). A History of Software Inspections. In Manfred Broy and Ernst Denert (Eds.), *Software Pioneers - Contributions to Software Engineering* (pp. 562–573). doi:10.1007/978-3-642-59412-0_34. 147

[7]   Colin M. Gray. (2013). Informal Peer Critique and the Negotiation of Habitus in a Design Studio. *Art, Design & Communication in Higher Education*, 12(2), 195–209. doi:10.1386/adch.12.2.195_1. 147

[8]   Colin M. Gray. (2018). Narrative Qualities of Design Argumentation. In Brad Hokanson, Gregory Clinton and Karen Kaminski (Eds.), *Educational Technology and Narrative* (pp. 51–64): Springer International Publishing. doi:10.1007/978-3-319-69914-1_5. 148

[9]   Karen Holtzblatt and Hugh Beyer. (2017). *Contextual Design: Design for Life*. Boston: Morgan Kaufmann. 167

[10]   Karen Holtzblatt, Jessamyn Burns Wendell, and Shelley Wood. (2005). *Rapid Contextual Design: A How-to Guide to Key Techniques for User-Centered Design*. Boston: Morgan Kaufmann. doi:10.1145/1066348.1066325. 167

[11]   Shane McIntosh, Yasutaka Kamei, Bram Adams, and Ahmed E. Hassan. (2014). The Impact of Code Review Coverage and Code Review Participation on Software Quality: A Case Study of the Qt, VTK, and ITK Projects. *MSR 2014: Proceedings of the 11th Working Conference on Mining Software Repositories*, 192–201. doi:10.1145/2597073.2597076. 148

[12]   Lucy Suchman. (1987). *Plans and Situated Actions: The Problem of Human-Machine Communication*. Cambridge: Cambridge University Press. 157

SNEAK ATTACKS ON KEY PROCESSES: AGILE

[1]   Michael Ahmadi, Anne Weibert, Rebecca Eilert, Volker Wulf, and Nicola Marsden. (2020). Feminist Living Labs as Research Infrastructures for HCI: The Case of a Video Game Company. *Proceedings of the 2020 CHI Conference on Human Factors in Computing Systems*, 1–15. doi:10.1145/3313831.3376716. 173

[2]   Michael Ahmadi, Anne Weibert, Corinna Ogonowski, Konstantin Aal, Kristian Gäckle, Nicola Marsden, and Volker Wulf. (2018). Challenges and Lessons Learned by Applying Living Labs in Gender and IT Contexts. *4th Gender & IT Conference (GenderIT'18)*, 239–249. doi:10.1145/3196839.3196878. 173

[3]   Asli Yüksel Aksekili and Christoph Johann Stettina. (2021). Women in Agile: The Impact of Organizational Support for Women's Advancement on Teamwork Quality and Performance in Agile Software Development Teams. *International Conference on Lean and Agile Software Development*, 3–23. doi:10.1007/978-3-030-67084-9_1. 172

[4]   Helena Barke. (2018). How Many Story Points for Diversity? Estimation as a Chance for Diversity Reflexion. *4th Gender & IT Conference (GenderIT'18)*, 221–223. doi:10.1145/3196839.3196872. 184, 186

[5]   Julia B. Bear, Lily Cushenbery, Manuel London, and Gary D. Sherman. (2017). Performance Feedback, Power Retention, and the Gender Gap in Leadership. *The Leadership Quarterly*, 28(6), 721–740. doi:10.1016/j.leaqua.2017.02.003. 178

[6]   Kent Beck, Mike Beedle, Arie van Bennekum, Alistair Cockburn, Ward Cunningham, Martin Fowler, James Grenning, Jim Highsmith, Andrew Hunt, Ron Jeffries, Jon Kern, Brian Marick, Robert C. Martin, Steve Mellor, Ken Schwaber, Jeff Sutherland, and Dave Thomas. (2001). Manifesto for Agile Software Development. Retrieved from http://agilemanifesto.org/. 170

[7]   Hugh Beyer. (2010). *User-Centered Agile Methods*. Morgan & Claypool Publishers. doi:10.2200/S00286ED1V01Y201002HCI010. 172

[8]   Ilya Bibik. (2018). *How to Kill the Scrum Monster - Quick Start to Agile Scrum Methodology and the Scrum Master Role*. Springer. doi:10.1007/978-1-4842-3691-8. 172

[9]   Iris Bohnet. (2016). *What Works: Gender Equality by Design*. Cambridge, MA: Harvard University Press. doi:10.4159/9780674545991. 180

[10]   Jutta Eckstein. (2019). *Retrospectives for Organizational Change: An Agile Approach*. Braunschweig, Germany. 172

[11]   Jutta Eckstein and John Buck. (2018). *Company-Wide Agility with Beyond Budgeting, Open Space & Sociocracy*. Leanpub. 172

[12]   Orit Hazzan and Yael Dubinsky. (2006). Empower Gender Diversity with Agile Software Development. In Eileen Trauth (Ed.), *The Encyclopedia of Gender and Information Technology* (pp. 249–256). Hershey, PA: IGI Global. doi:10.4018/978-1-59140-815-4.ch039. 172

[13]   Seulki Jang, Tammy D. Allen, and Joseph Regina. (2020). Office Housework, Burnout, and Promotion: Does Gender Matter? *Journal of Business and Psychology*, 36, 793–805. doi:10.1007/s10869-020-09706-3. 186

[14]   Ken H Judy. (2012). Agile Values, Innovation and the Shortage of Women Software Developers. In *System Science (HICSS), 2012 45th Hawaii International Conference on IEEE*, 5279–5288. doi:10.1109/HICSS.2012.92. 172

[15]   Desmond J. Leach, Steven G. Rogelberg, Peter B. Warr, and Jennifer L. Burnfield. (2009). Perceived Meeting Effectiveness: The Role of Design Characteristics. *Journal of Business and Psychology*, 24(1), 65–76. doi:10.1007/s10869-009-9092-6. 180

[16]   Janice Fanning Madden. (2012). Performance-Support Bias and the Gender Pay Gap among Stockbrokers. *Gender & Society*, 26(3), 488–518. doi:10.1177/0891243212438546. 186

[17]   Nicola Marsden, Michael Ahmadi, Volker Wulf, and Karen Holtzblatt. (2021). Surfacing Challenges in Scrum for Women in Tech. *IEEE Software*, 1–8. doi:10.1109/MS.2021.3115461. 169

[18]   Zainab Masood, Rashina Hoda, and Kelly Blincoe. (2020). How Agile Teams Make Self-Assignment Work: A Grounded Theory Study. *Empirical Software Engineering*, 25(6), 4962–5005. doi:10.1007/s10664-020-09876-x. 186

[19]   Bertrand Meyer. (2014). *Agile! The Good, the Hype and the Ugly*. Switzerland: Springer International Publishing. 170, 172, 183

[20]   Nils Brede Moe, Torgeir Dingsøyr, and Tore Dybå. (2010). A Teamwork Model for Understanding an Agile Team: A Case Study of a Scrum Project. *Information and Software Technology*, 52(5), 480–491. doi:10.1016/j.infsof.2009.11.004. 178

[21]   Serge Moscovici and Claude Faucheux. (1972). Social Influence, Conformity Bias, and the Study of Active Minorities. *Advances in Experimental Social Psychology*, 6, 149–202. doi:10.1016/S0065-2601(08)60027-1. 178

[22]   Valerie Purdie-Vaughns and Richard P. Eibach. (2008). Intersectional Invisibility: The Distinctive Advantages and Disadvantages of Multiple Subordinate-Group Identities. *Sex Roles*, 59(5–6), 377–391. doi:10.1007/s11199-008-9424-4. 178

[23]   Nancy Russo. (2015). Making IT Matter for Women: Exploring Agile Perceptions. *UK Academy for Information Systems Conference Proceedings*, 18, 1–10. 172

[24]   Sachverständigenkommission. (2021). Digitalisierung geschlechtergerecht gestalten - Gutachten für den Dritten Gleichstellungsbericht der Bundesregierung. Retrieved from https://www.dritter-gleichstellungsbericht.de/de/topic/73.gutachten.html. 179

[25]   Sonja A. Sackmann. (1991). Uncovering Culture in Organizations. *The Journal of Applied Behavioral Science*, 27(3), 295–317. doi:10.1177/0021886391273005. 173

[26]   Ken Schwaber and Jeff Sutherland. (2012). *Software in 30 Days: How Agile Managers Beat the Odds, Delight Their Customers, and Leave Competitors in the Dust*. John Wiley & Sons. doi:10.1002/9781119203278. 178

[27]   Ken Schwaber and Jeff Sutherland. (2020). The Scrum Guide. Retrieved from https://scrumguides.org/docs/scrumguide/v2020/2020-Scrum-Guide-US.pdf. 172

[28]   Hirotaka Takeuchi and Ikujiro Nonaka. (1986). The New New Product Development Game. *Harvard Business Review*, 64(1), 137–146. 178

[29]   VersionOne. (2020). 14th Annual State of Agile Report. Retrieved from https://stateofagile.com/ - ufh-i-615706098-14th-annual-state-of-agile-report/7027494. 170

[30]   Joan C. Williams and Sky Mihaylo. (2019). How the Best Bosses Interrupt Bias on Their Teams. *Harvard Business Review*, Nov./Dec. 169

[31]   Joan C. Williams, Katherine W. Phillips, and Erika V. Hall. (2016). Tools for Change: Boosting the Retention of Women in the STEM Pipeline. *Journal of Research in Gender Studies*, 6(1), 11–75. doi:10.22381/JRGS6120161. 178

VALUING AND JERK BEHAVIORS

[1] Laura Batista and Thomas G. Reio Jr. (2019). Occupational Stress and Instigator Workplace Incivility as Moderated by Personality: A Test of an Occupational Stress and Workplace Incivility Model. *Journal of Organizational Psychology*, 19(2), 38–49. doi:10.33423/jop.v19i2.2042. 197

[2] Katy Cook. (2020). *The Psychology of Silicon Valley*. Springer. doi:10.1007/978-3-030-27364-4. 191

[3] Kimberly Earles. (2020). The Gender Divide in the Tech Sector. Retrieved from https://georgetown.southseattle.edu/sites/georgetown.southseattle.edu/files/2020-10/Gender-DivideInTech-Final%5B5146%5D_1.pdf. 197

[4] Lucia Happe and Barbora Buhnova. (2021). Frustrations Steering Women Away from Tech. *IEEE Software*, 1–6. doi:10.1109/MS.2021.3099077. 191

[5] Constance K. Jensen. (2018). *The Essence of Experiencing Gender Disparity by Women in Tech*. The University of the Rockies. 191

[6] Etlyn J. Kenny and Rory Donnelly. (2020). Navigating the Gender Structure in Information Technology: How Does This Affect the Experiences and Behaviours of Women? *Human Relations*, 73(3), 326–350. doi:10.1177/0018726719828449. 191

[7] Pauline Schilpzand, Irene E. De Pater, and Amir Erez. (2016). Workplace Incivility: A Review of the Literature and Agenda for Future Research. *Journal of Organizational behavior*, 37(S1), S57–S88. doi:10.1002/job.1976. 197

[8] Trustradius. (2021). Women in Tech Report. Retrieved from https://www.trustradius.com/buyer-blog/women-in-tech-report. 191

BUILDING RESILIENCE: TEAM MANIFESTO AND PROCESS CHECKS

[1] Clint Bowers, Christine Kreutzer, Janis Cannon-Bowers, and Jerry Lamb. (2017). Team Resilience as a Second-Order Emergent State: A Theoretical Model and Research Directions. *Frontiers in Psychology*, 8, 1360. doi:10.3389/fpsyg.2017.01360. 211

[2] John T. Byrd and Michael R. Luthy. (2010). Improving Group Dynamics: Creating a Team Charter. *Academy of Educational Leadership Journal*, 14(1), 13–26. 212

[3] Shelley Correll and Caroline Simard. (2016). Vague Feedback Is Holding Women Back. *Harvard Business Review*, 94, 2–5. 217

[4]   William Y. Degbey and Katja Einola. (2020). Resilience in Virtual Teams: Developing the Capacity to Bounce Back. *Applied Psychology*, 1-37. doi:10.1111/apps.12220. 211

[5]   Lucas Gren, Alfredo Goldman, and Christian Jacobsson. (2020). Agile Ways of Working: A Team Maturity Perspective. *Journal of Software: Evolution and Process*, 32(6), e2244. doi:10.1002/smr.2244. 216

[6]   Lily Jampol and Vivian Zayas. (2021). Gendered White Lies: Women Are Given Inflated Performance Feedback Compared with Men. *Personality and Social Psychology Bulletin*, 47(1), 57–69. doi:10.1177/0146167220916622. 217

[7]   John E. Mathieu and Tammy L. Rapp. (2009). Laying the Foundation for Successful Team Performance Trajectories: The Roles of Team Charters and Performance Strategies. *Journal of Applied Psychology*, 94(1), 90–103. doi:10.1037/a0013257. 212

[8]   Barry Overeem. (2014). The Team Manifesto; the Foundation Every Team Needs. https://www.linkedin.com/pulse/team-manifesto-foundation-every-barry. 212

[9]   Elissa L. Perry, Caryn J. Block, and Debra A. Noumair. (2020). Leading In: Inclusive Leadership, Inclusive Climates and Sexual Harassment. *Equality, Diversity and Inclusion: An International Journal*. 40(4), 430–447. doi:10.1108/EDI-04-2019-0120. 212

[10]   Denise Salin. (2020). 'Competent' or 'Considerate'? The Persistence of Gender Bias in Evaluation of Leaders. *Nordic Journal of Working Life Studies*. 10(1), 59–79. doi:10.18291/njwls.v10i1.118680. 217

[11]   Ken Schwaber and Jeff Sutherland. (2020). The Scrum Guide. Retrieved from https://scrumguides.org/docs/scrumguide/v2020/2020-Scrum-Guide-US.pdf. 216

[12]   Vanessa Sequeira. (2021). Miroverse Template: Team Charter. Retrieved from https://miro.com/miroverse/team-charter/. 212

[13]   Rachel Emma Silverman. (2015). Gender Bias at Work Turns up in Feedback. *The Wall Street Journal*. Retrieved from https://www.wsj.com/articles/gender-bias-at-work-turns-up-in-feedback-1443600759. 217

[14]   Therese E. Sverdrup and Vidar Schei. (2015). "Cut Me Some Slack" the Psychological Contracts as a Foundation for Understanding Team Charters. *The Journal of Applied Behavioral Science*, 51(4), 451–478. doi:10.1177/0021886314566075. 211

[15]   Therese E. Sverdrup, Vidar Schei, and Øystein A. Tjølsen. (2017). Expecting the Unexpected: Using Team Charters to Handle Disruptions and Facilitate Team Performance. *Group Dynamics: Theory, Research, and Practice*, 21(1), 53–59. doi:10.1037/gdn0000059. 212

CONCLUSION: PRINCIPLES OF PROCESS INTERVENTION FOR RETAINING WOMEN IN TECH

[1] Isabel Bilotta, Shannon K. Cheng, Linnea C. Ng, Abby R. Corrington, Ivy Watson, Jensine Paoletti, Mikki R. Hebl, and Eden B. King. (2021). Remote Communication Amid the Coronavirus Pandemic: Optimizing Interpersonal Dynamics and Team Performance. *Industrial and Organizational Psychology*, 14(1–2), 36–40. doi:10.1017/iop.2021.10. 228

[2] Larry Dignan. (2019). Meetings Suck and Bad Ones Will Cost Enterprises $399 Billion in 2019, Says Doodle. Retrieved from https://www.zdnet.com/article/meetings-suck-and-bad-ones-will-cost-enterprises-399-billion-in-2019-says-doodle/. 236

[3] Laura M. Giurge, Ashley V. Whillans, and Ayse Yemiscigil. (2021). A Multicountry Perspective on Gender Differences in Time Use During Covid-19. *Proceedings of the National Academy of Sciences*, 118(12), 1–7. doi: 10.1073/pnas.2018494118. 230

[4] Alisha Haridasani Gupta. (2020, April 14). It's Not Just You: In Online Meetings, Many Women Can't Get a Word In. In *The New York Times*. 228

[5] Katherine A. Karl, Joy V. Peluchette, and Navid Aghakhani. (2021). Virtual Work Meetings During the Covid-19 Pandemic: The Good, Bad, and Ugly. *Small Group Research*, 1–23. doi:10.1177/10464964211015286. 228

[6] Paige Rodeghero, Thomas Zimmermann, Brian Houck, and Denae Ford. (2020). Please Turn Your Cameras On: Remote Onboarding of Software Developers During a Pandemic. *IEEE/ACM 43rd International Conference on Software Engineering: Software Engineering in Practice (ICSE-SEIP)*, 1–10. doi:10.1109/ICSE-SEIP52600.2021.00013. 231

[7] Ingo Siegert and Oliver Niebuhr. (2021). Case Report: Women, Be Aware That Your Vocal Charisma Can Dwindle in Remote Meetings. *Frontiers in Communication*, 5(Jan.), 1–7. doi:10.3389/fcomm.2020.611555. 231

[8] Maria Tomprou, Young Ji Kim, Prerna Chikersal, Anita Williams Woolley, and Laura A. Dabbish. (2021). Speaking out of Turn: How Video Conferencing Reduces Vocal Synchrony and Collective Intelligence. *PloS one*, 16(3), 1–14. doi:10.1371/journal.pone.0247655. 230

[9] Vanessa Wasche. (2021). Women, Don't Turn Your Camera Off for Zoom Calls. Fast Company. Retrieved from https://www.fastcompany.com/90577160/women-dont-turn-your-camera-off-for-zoom-calls. 231

Authors' Biographies

Karen Holtzblatt is a thought leader, industry speaker, and author. As co-founder and CEO of InContext Design, Karen is the visionary behind Contextual Design, a user-centered design approach used by universities and companies worldwide. Her latest book, *Contextual Design, 2nd Edition: Design for Life*, details the new techniques needed for product innovation today. Recognized as a leader in requirements and design, Karen has been twice honored by the ACM SIGCHI (Special Interest Group on Computer-Human Interaction), an international society for professionals, academics, and students. She was first awarded membership to the CHI Academy and then received the first Lifetime Award for Practice presented in recognition of her impact on the field.

Karen is also the Executive Director of WITops, a research wing of InContext dedicated to finding practical and effective ways to understand and influence the experience of diverse people working in tech companies, especially women. Known for pioneering transformative approaches, in 2013 Karen turned her energy to the challenges of retaining women in technology. Through collaboration with volunteer colleagues and students worldwide, WITops teams have developed the @Work Experience Framework and Measure defining the key workplace experiences needed for women to thrive. Then they turned their attention to solutions, developing critical intervention techniques and materials to help companies improve their culture for diverse teams. All work at WITops is grounded in deep field research and quantitative techniques which drive ideation and iteration of tested solutions.

Karen is a Research Scientist at the University of Maryland's iSchool. She regularly consults with universities to help improve their HCI programs, including building in awareness of issues that new workers, especially diverse people, will face. Karen has more than 30 years of teaching experience, professionally and in university settings. Karen holds a doctorate in applied psychology from the University of Toronto. Contact Karen with questions at karen@incontextdesign.com.

Nicola Marsden is a professor of social informatics at Heilbronn University, Germany. She combines insights from psychology, software engineering, design research, and organizational behavior to improve collaboration and foster innovation in technology development. Her research is based on a combination of experience in both academia and industry, often with a gender or cross-cultural perspective. In her transformation work with people, teams, and organizations she offers a balance between scientific knowledge and a practical approach.

Throughout her career, Nicola has worked toward bias-free equal opportunity for women. Nicola is vice-chair of the Competence Center Technology, Diversity, Equal Opportunities in Germany (kompetenzz.de), a non-profit dedicated to ensuring equal opportunity for women and men in STEM education and industry. The competence center produces research-based resources and developed and coordinates nationwide initiatives such as the Girls' Day or Boys' Day program. This program helps young people choose a profession based on their individual strengths and talents, rather than clichés or gender stereotypes.

Nicola's most recent academic research examines behavioral design to de-bias collaboration, human–computer interaction, and design processes. She has also worked as a key collaborator with Karen Holtzblatt on the mission of WITops to understand and create solutions to retain women in technology.

Her extensive work with corporations also uses a theory-based, practical systems perspective to design, implement and manage innovation projects, change processes, training and development programs, and strategic development projects. Her long-standing experience working with organizations in different countries allows Nicola to translate research into everyday practices to improve group dynamics and perspectives. Nicola holds a doctorate in social psychology from Saarland University. Contact Nicola with questions at: nicola.marsden@hs-heilbronn.de.